HERITAGE
APPLES
A NEW SENSATION
SUSAN LUNDY

TouchWood
Editions

TouchWood Editions
touchwoodeditions.com

LIBRARY AND ARCHIVES CANADA CATALOGUING IN PUBLICATION
Lundy, Susan, 1965–
Heritage apples : a new sensation / Susan Lundy.

Includes index.
Issued also in electronic formats.
ISBN 978-1-927129-91-3

1. Apples. 2. Apples—Varieties. 3. Cooking (Apples). I. Title.

SB363.L85 2013   634'.11   C2012-907742-9

Editor: Marlyn Horsdal
Proofreader: Holland Gidney
Design: Pete Kohut
Cover image: Derrick Lundy
Author photo: Sierra Lundy
All photos courtesy the author unless otherwise indicated.

We gratefully acknowledge the financial support for our publishing activities from the Government of Canada through the Canada Book Fund, Canada Council for the Arts, and the province of British Columbia through the British Columbia Arts Council and the Book Publishing Tax Credit.

MIX
Paper from
responsible sources
FSC® C016973

This book was produced using FSC®-certified, acid-free paper, processed chlorine free and printed with soya-based inks.

The information in this book is true and complete to the best of the author's knowledge. All recommendations are made without guarantee on the part of the author. The author disclaims any liability in connection with the use of this information.

1 2 3 4 5 17 16 15 14 13

PRINTED IN CHINA

Bags of heritage apples to my daughters, Danica (Newton Pippins for cider) and Sierra (Sierra Beautys, because she is one); to my mother, Barbara (Wolf Rivers for pie); and to Bruce (Cox's Orange Pippins, from our trees, I hope).

CONTENTS

CHAPTER ONE
The Apples of My Eye

Baskets and boxes of heritage apples for sale at Ruckle Farm on Salt Spring Island. Considered the oldest working farm in British Columbia, it has heritage apple trees that have been producing fruit for over 140 years.

Crates of Cox's Orange Pippins, grown at Heart Achers Farm in Cawston, BC, are set for shipment via a fruit packing company called Direct Organics Plus.

Apple Facts

The common apple tree is a member of the Rosaceae family, which originated in western Asia and has been cultivated for more than four thousand years. There are more than seventy-five hundred varieties of apples worldwide.

Apples account for 50 per cent of deciduous fruit tree production in the world.

One of the world's most famous apples—from the Bible's Garden of Eden—may not have been an apple at all. It's possible that Christian scholars took the story's "forbidden fruit" to be an apple because the Latin word *malum* means both "apple" and "evil." The forbidden fruit was more likely a fig (because the next verse talks about sewing fig leaves into loincloths), or a pomegranate, which was native to the region.

On April 1, 1976, Steve Wozniak and Steve Jobs started Apple Computers—so named, according to several sources, because Jobs in particular was a health-conscious natural foods advocate. In 1981, Jef Raskin, an Apple employee, got permission to "build his own dream computer. He wanted it to be inexpensive, portable, and as easy to use as an appliance. He called it Macintosh [*sic*] after his favorite kind of apple."[1]

When I announced to people, "I'm writing a book about heritage apples," they stared at me blankly for an instant. Then a gleam appeared in their eyes and they asked, "What is a heritage apple, anyway?" Before long, they were enthusiastically mining their memories, telling stories of apples recalled from childhood, and seemingly smelling the scent of the ripening fruit in their grandparents' backyards all over again.

My own journey into the world of heritage apples began with this book. I Googled "heritage apples." I drank cider made from heritage apples. I travelled to heritage apple orchards, talked with apple people, attended apple festivals and apple-growing lectures. Apples started popping up everywhere. At a birthday cocktail party, I mentioned to the host that I was working on this book. His eyes lit up, he waved toward the half-dozen old-variety trees on his property (including Golden Delicious, Gala, and King) and, practically bursting with enthusiasm, ushered me into the basement, displaying his apple cider-making equipment. Back upstairs, he said, was a guest who makes apple brandy from his apples. That got my mouth watering and I spent the rest of the party talking apples.

I found myself drawn to apple products: checking ingredients on apple cider labels; stopping to smell the fragrance of apple bath products and lotions; ordering a cocktail called "Apple" (apple-flavoured vodka, apple Sour Puss, Sprite, and cinnamon—delicious!). Then it got closer to home. It turns out I have a heritage apple tree on my five-acre property on Salt Spring Island in British Columbia. It doesn't look like an apple tree, as the tiny, yellow crabapples are smaller than the fruit on my cherry tree out back. I took the apples to be identified, but no one was able to name them—something that's not too surprising, given the nature of apple reproduction.

Here's the thing with apples—you can't just plant a seed from an apple and have the same variety grow in its spot. In the same way human offspring are genetically unique—with DNA from both parents—so it is with apples. In fact, if you took all the seeds in one apple, say five of them, you'd get five different, unique apple varieties. To reproduce a specific apple tree, you need to employ the age-old practice of grafting: taking a piece of the original tree (the scion) and inserting it into rootstock, so the tissues grow together and produce a clone of the original tree.

There isn't a straight answer to "What is a heritage apple?" But in general, people consider apple varieties originating prior to 1950 as heritage. Several common apples found in grocery stores—McIntosh, Red Delicious, Golden Delicious—are heritage varieties, albeit dramatically changed in taste and appearance through years of domestic breeding. Others, like Fuji (Japan, 1962), Ginger Gold (Virginia, 1982), and Ambrosia (BC, 1987), are newer apples developed specifically for the market. Outside of grocery store bins, there are thousands of apple varieties—at least seventy-five hundred named apple cultivars worldwide. Between four hundred and five hundred types are available through nurseries, while only a dozen or so can be found in grocery stores. This leaves a lot of room for apple treasure hunting—something more and more people are discovering—because, as it turns out, all these apples have different flavours and qualities, some subtle, some dramatic.

The next "close-to-home" apple surprise came during Salt Spring's annual apple festival as I visited Ruckle Farm, a heritage site, which has dozens of 140-year-old apple trees still producing fruit. I went there with my ex-husband, a photographer, hoping he could get some images for this book. As we toured, he started talking about the trees, pointing out characteristics that differentiate heritage apple trees from newer trees, reeling off growing information, discussing disease and pruning techniques. I was speechless.

"How do you know all this?" I demanded of the man to whom I was married for sixteen years. Turns out that in the 1970s, he was in line to take over an orchard in Kamloops and spent months preparing for the job. He knew a lot about apples, which proved helpful because I suddenly had someone I could ply with stupid questions. (I mean, what did it matter if he thought I was stupid; we were already divorced anyway.)

The fact that I lived on Salt Spring was a bit of heritage apple happenstance as well. In the late 1800s, Salt Spring was the major fruit-producing area in BC. The first apple tree was probably planted there in 1858 by Henry Trage, and by 1894, the island's 4,600 fruit trees outnumbered residents ten to one. Some of those trees, now 150 years old, still exist on Salt Spring (including some original Trage trees at Wave Hill Farm), although the large orchards of 1,000-plus trees are gone.

These tiny, yellow crabapples were discovered growing on a heritage apple tree in the author's front yard on Salt Spring. No one has been able to identify the apples, which are smaller than the fruit on a nearby cherry tree.

But recently there has been a revival of that apple-growing tradition, with a trend toward cultivating heritage varieties. In fact, about 350 known apple types are being grown on Salt Spring. Much of this has to do with people like Dr. Bob Weeden, who grows and sells nothing but heritage varieties at an orchard he planted himself 20 years ago, and, of course, Harry Burton (AKA Captain Apple), owner of Apple Luscious Organic Orchard.

Google "heritage apples," and Harry almost always comes up. Even unlikely places, like the Creemore Heritage Apple Society—on the other side of the country from Harry—include him in "links and references" sections. Harry is the "go to" man for apples. How many times have I heard a sentence that starts something like, "Yeah, had this big old King tree that looked like it was gonna come down in the wind, so I called Harry."

A professor of environmental protection for twenty years in Ontario before moving out west, Harry planted his five-acre Salt Spring orchard in 1986 on land that had been logged in 1980. Harry has about three hundred trees, mostly apples but some plums, pears, and cherries as well. His focus is growing the "best tasting connoisseur apples in the world, both heritage and new, as well as red-fleshed apples."

In addition to growing apples, grafting apples, trying out new apple varieties, and selling apples and apple trees, Harry has become a legend on Salt Spring, where he has organized an apple festival since the late 1990s, drawing an ever-larger crowd of people and spreading the word on everything apple. Harry may appear apple-crazy (well, he is), but he is also apple-knowledgeable. The apple festival includes orchard tours and a cornucopia of apple education: factoids, history, anecdotes. You name it about apples, Harry probably knows it.

So in my pursuit of heritage apples, I met a lot of apple people without even stepping far from home. But as I did step away—meeting heritage apple people in BC and Alberta, communicating via email and phone with farmers elsewhere, and reading articles and checking websites online—I discovered heritage apples are everywhere. In fact, I began to feel a little thrill of excitement, as though we heritage apple people (notice, I now included myself in this group) were on the edge of something that could be big. Slowly but steadily, interest in heritage apples was growing alongside other food movements, especially those that centred around organic and locally produced food.

As food activist Michael Pollan has noted, current food issues abound (ranging from animal rights to campaigns against genetically modified crops), but at its base, "The food movement is also about community, identity, pleasure, and, most notably, about carving out a new social and economic space removed from the influence of big corporations on the one side and government on the other."[2]

Historically, the demise of many varieties of heritage apples occurred alongside the growth of big apple-producing commercial ventures, which sought apples that met consumer expectations in the areas of appearance, taste, and a longer shelf life. Apples with brown russeting (a brownish colour found on the skin of the apple), for example, were out, and big, red, shiny apples were in. Tart or bitter apples were out; apples developed to match our urge for sweetness were in.

But as Pollan noted, "The attempt to redefine . . . the traditional role of the consumer has become an important aspiration of the food movement . . . The modern marketplace would have us decide what to buy strictly on the basis of price and self-interest; the food movement implicitly proposes that we enlarge our understanding of both those terms, suggesting that not just 'good value' but ethical and political values should inform our buying decisions, and that we'll get more satisfaction from our eating when they do." [3]

A resurgence in heritage apples fits nicely into this package. Since there are so many apple tastes and varieties out there, why stick with the dozen or so in grocery store bins? Why not buy apples from small, local producers and know exactly where our food is coming from? In my own personal journey into the world of heritage apples, I've gone from throwing a few Braeburns and Fujis into a bag at the store, to looking for more unique apples. I don't want the same old commercial varieties because I now know how much potential is out there. My apple world has not only expanded, it has exploded.

I tasted more apple varieties in six months of researching this book than in my entire life and I continued to be stunned (and delighted) to discover the multitude of different flavours. Never could I have imagined that a tasting event with a hundred different apples spread out on a table could result in so many variations of taste. And yes, after tasting dozens of different apples, I settled on a favourite—although I know there are so many others out there that this could change at any time. My personal top-tasting old apple is not some obscure variety found growing in the wilds of Kazakhstan, where recent DNA studies confirmed that the cultivated apple originated. No, my dream apple, Cox's Orange Pippin, discovered in 1825, is hugely popular, especially in the United Kingdom. But it is also delicious, with a subtle citrus flavour. It was originally grown in Buckinghamshire, England, by a retired brewer and horticulturalist, Richard Cox, and is the parent of numerous other popular apples.

While taste and food politics may spur interest in old apple varieties, there is another factor leading to a revival as noted by Ron Schneider at Heart Achers Farm in Cawston, BC. He took a risk several years ago, planting a number of heritage apple trees mostly aimed at the cider-making market. In the past couple of years, he says, interest has boomed, with people coming by his orchard to check out his trees and pick his brain. His production of apples for a cidery near Victoria exploded to 150,000 pounds of apples in 2011, paralleling the company's leap from selling 10,000 litres of cider a year to 40,000 litres. Schneider has been visited by others interested in establishing cider apple orchards, using heritage cider apples, and also says there is a growing interest on the "fresh" market for old varieties such as Newton Pippin and Winesap.

Interest in apple varieties can certainly be seen at the myriad apple festivals held in all parts of the country. Again, I was stunned by the sheer number of apple-frenzied hordes at an apple festival in the University of British Columbia's botanical garden. Since its

Ron Schneider at Heart Achers Farm in Cawston, BC, holds one of his award-winning Cox's Orange Pippins as he discusses his decision to go against the flow and grow heritage apples.

humble beginnings twenty years ago, the festival has blossomed into a major two-day event, attended by fourteen thousand people.

There I met up with Clay Whitney, a landscaper who has taken it upon himself to save clones of heritage apple trees falling to development in Sooke on Vancouver Island. An expert in apple identification, Clay gave me a memorable lesson in apple diversity. Apples, it turned out, are more than skin, flesh, stems, and cores. They are carpels, suture lines, and lenticils. They have shoulders, basins, bases, and apex. They can be any variation of red, yellow, green, or russeted.

Lots of heritage apples have fascinating histories, too. For example, Canada's most famous apple, the McIntosh, was discovered by John McIntosh in 1811. Through further "apple happenstance," I met a descendant of McIntosh's at a party in Calgary. Lynne Walker is the great-granddaughter of Hester McIntosh, daughter of Allan, one of John McIntosh's thirteen or fourteen children. Hester married Lynne's great-grandfather, Richard Kendrick, in 1872, and Lynne grew up in the house they eventually built in Manitoba.

According to Lynne, who has done extensive research on her famous ancestor, John McIntosh was born in the Mohawk Valley, in upstate New York, in 1777. His parents had immigrated to America from Inverness, Scotland, earlier that year.

"John was a United Empire Loyalist and came to Canada after the American Revolution, spending his first few years living on the banks of the St. Lawrence River. He settled on a farm in Matilda Township, near a settlement once known as 'McIntosh's Corners,' now called Dundela, Ontario, and married Hannah Doran in 1801," Lynne said.

OPPOSITE: One of the 140-year-old apple trees, as well as an old barn, at Ruckle Farm on Salt Spring Island. The working farm is now part of a 1,200-acre park, and the orchards and old buildings are open to the public for viewing.
PHOTO: DERRICK LUNDY

"While clearing the land, he came across several apple trees, [which] may have come from the seeds of two or three different apples, possibly Fameuse. Those original seedlings flourished for a time, but by 1830, only one remained. This became the McIntosh Red, so named because of the family's name and the distinctive colour of the skin. John did not realize the worth of his tree but his son, Allan, grafted and budded scions of that tree onto other varieties of apple trees, and established a nursery. The [original] tree was situated only about fifteen feet from the house and was badly damaged when the house caught fire in 1894. However, it continued to bear until about 1906, when the leaves began to wither and the tree gradually died."

Lynne said the residents of the area erected a monument with the following inscription: "The original McIntosh apple tree stood about twenty rods north of this spot. It was one of a number of seedlings taken from the border of the clearing and transplanted by John McIntosh in 1796. Erected by popular subscription in 1912." (There is an error in the date inscribed, Lynne pointed out, as John did not settle in the area until the late 1790s.)

The McIntosh apple made headlines again in July 2011, the year several communities celebrated the 200th anniversary of the McIntosh. It was then that the last-known, first-generation graft taken from the original tree died and was cut down; it was 150 years old. The tree had been grown from cuttings taken from the original tree and planted by Samuel Smyth, who worked for McIntosh. Horticulturalists took a dozen twigs from Smyth's tree and grafted them onto rootstock, thus continuing the lineage of the first tree.

When I asked Lynne what her favourite kind of apple is, she said, "Even before I made the family connection to the McIntosh, my favourite apple has always been the Mac. Just something about the sweet flesh—I find other apples 'dry' by comparison."

So in the course of a few months, I'd gone from being able to safely distinguish an apple from an orange, but not much more, to having this wealth of information—and a new passion for heritage varieties. One night, I said to my partner, Bruce, "If I start talking about apples in the bar tonight, please kick me under the table." But then I couldn't resist a comment or two. Because it was all so fascinating!

Here's what Harry Burton had to say on the subject, "I feel very lucky to be an avid apple grower. What a beautiful direction to channel your efforts! A further reward is available for those growers seeking new, heritage, or unique varieties of apples. Then it becomes a passion. You're lost and there is no turning back."

COX'S ORANGE PIPPIN

ENGLAND, 1825[4]

Taste and appearance: Delicious, sweet, and slightly acidic. Red-orange flush with cream-coloured flesh.

Use: Eating fresh, juice, cooking, and hard cider.

History: Originally grown in Buckinghamshire, England, in 1825 by retired brewer and horticulturalist Richard Cox from the seed of a Ribston Pippin.

Growing and harvesting: Trees are fairly productive but susceptible to scab and canker, and need to be grown in sunny, well-drained areas. Apples are picked in late September and keep about four months.

Other: Highly popular in Europe. A number of crosses and sports have been discovered over subsequent years, retaining "Cox" in their names (e.g., Crimson Cox, King Cox, Queen Cox).

PHOTO: CLAY WHITNEY

Taste and appearance: Very sweet. A yellow, gold-coloured apple.

Use: Eating fresh and cooking. It is popular in salads, applesauce, and apple butter.

History: Discovered as a chance seedling, possibly of Grimes Golden, on the Mullins family farm in West Virginia. According to Wikipedia, it was known locally as the Mullins Yellow Seedling and Annit apple. Anderson Mullins sold the tree and propagation rights to Stark Brothers Nurseries, which paid five thousand dollars for it, and kept it in a wire cage for years. The apple was first marketed in 1914 by Stark Brothers as a companion to Red Delicious, although the two apples are not related.

Growing and harvesting: Prone to bruising and shrivelling, it needs to be handled and stored carefully. Trees need a long growing season; the apple is picked mid- to late October and keeps well in storage until June.

Other: Because it is so sweet, cooks say they use less sugar when recipes call for Golden Delicious. The flesh is crisp, firm, juicy, and holds its shape when cooked.

McINTOSH
ONTARIO, 1811

PHOTO: CLAY WHITNEY

Taste and appearance: Medium-sized and round, with a bright red colour and green or yellow splashes on one side. The trademark white flesh is characteristically aromatic, juicy, and tender, becoming mild and nearly sweet when ripe.

Use: Eating fresh and cooking. (It retains its flavour as it's cooked, making it good in pies and applesauce.)

History: Discovered by John McIntosh in 1811 on his farm in Dundela, Ontario.

Growing and harvesting: It grows most successfully in climates with clear days and cool nights in the fall. It is picked mid-September and keeps until January, but is in its prime soon after picking.

Other: This apple, thought to be an offspring of Fameuse, is one of North America's most popular apples.

PHOTO: CLAY WHITNEY

RED DELICIOUS
IOWA, 1870s

Taste and appearance: Tall and wasp-waisted, it has five distinct knobs on the bottom; the flesh is yellow, juicy, and fine-grained with a mildly aromatic, sweet, mellow flavour when allowed to ripen fully.

Use: Eating fresh. (It does not hold its flavour when cooked.)

History: In 1872, Jesse Hiatt, a farmer in Peru, Iowa, discovered a sprout growing from the roots of a Bellflower seedling that had previously been cut down twice. As the story goes, he said, "If thee must live, thee may," and so let it grow. He found that it produced a striped, reddish-yellow conical fruit, which he called Hawkeye and entered in a contest held by Stark Brothers Nurseries to find an apple to replace Ben Davis. Stark Brothers bought the rights from Hiatt, and renamed it Red Delicious.

Growing and harvesting: Suitable for growing in warm climates, it's picked in early October and keeps six months.

Other: According to Wikipedia, today's version of Red Delicious—developed for its intensely red colour and early harvestability—has replaced the original cultivar in commercial orchards, and "the taste and texture of the harvested commodity have deteriorated, [with] many customers [now rejecting] Red Delicious at the food market."

Chickpea-Apple Curry

1 Tbsp extra-virgin olive oil

1 Tbsp curry powder

1 large onion, cut into ¼-inch-wide slivers

2 Tbsp grated fresh ginger

4 garlic cloves, minced

½ cup chopped fresh cilantro, plus more for garnish

1 28-oz can whole tomatoes

2 apples, peeled, cored, and cut into ½-inch pieces

2 15½-oz cans chickpeas, rinsed, drained

1 Tbsp freshly squeezed lemon juice

1 cup plain nonfat yogourt

Heat olive oil in a large saucepan over medium-low heat. Add curry powder; cook until fragrant, about 1 minute. Add onion, ginger, and garlic; cook until onion starts to soften and is well coated with curry mixture—2 to 3 minutes. Add cilantro, tomatoes, apples, and chickpeas. Cover; simmer until apples are tender, 30 to 40 minutes. Uncover; simmer until slightly thickened, about 5 minutes. Remove from heat, stir in lemon juice. Stir in yogourt just prior to serving.

Recipe provided by Steve Glavicich, chef/owner, Braizen Food Truck, Calgary

Chicken-Apple Pâté

1 lb chicken livers

½ lb skinless, boneless chicken breasts

1 small onion, halved

2 eggs

¼ cup half-and-half cream

¼ cup chicken broth

1 Tbsp cognac

1 tsp salt

2 stems fresh thyme

½ tsp nutmeg

1 cup shredded apple

bread or crackers, for serving

Preheat oven to 350 degrees. Lightly grease a loaf pan. Place chicken livers, breasts, and onion in a food processor; purée until coarsely ground. Add eggs, half-and-half, broth, cognac, salt, thyme, and nutmeg; purée until well blended. Stir in apple. Transfer mixture to prepared loaf pan. Cover tightly and bake for 75 minutes. Allow to cool for 1 hour at room temperature; then refrigerate for 2 hours. Unmold onto serving platter and serve with bread or crackers.

Recipe provided by Steve Glavicich, chef/owner, Braizen Food Truck, Calgary

The sight and scent of heritage apples often brings out an emotional response in people, who suddenly recall the apples of their youth.
PHOTO: DERRICK LUNDY

CHAPTER TWO
"A" is for Apple

Apple Facts

In 2010, an Italian-led consortium announced it had decoded the complete genome of the apple (Golden Delicious). It had about fifty-seven thousand genes, the highest number of any plant genome studied to date and more genes than the human genome (about thirty thousand).[1]

The old saying, "An apple a day, keeps the doctor away," may come from an old English adage, "To eat an apple before going to bed, will make the doctor beg his bread."[2]

Apple seeds are mildly poisonous, containing a small amount of amygdalin, a cyanogenic glycoside.

In Canada, major apple production areas include British Columbia, Ontario, Quebec, New Brunswick, and Nova Scotia. Approximately 60 per cent of apples produced in Canada are sold as fresh fruit. The remainder is used for juice, cider, pie filling, and baked goods.[3]

It happened again as I flipped through photographs from the Salt Spring Apple Festival and spotted an image which, taken from a balcony, looked down over a horde of people crowding around apple display tables. The tables, organized in a long circle, had "apple experts" available for questions on the inside. I looked at those experts, with a bit of apple-knowledge envy, and saw a familiar figure. Daphne? My good friend Daphne? Fellow soccer-travel mom? Fellow high-school-sports-on-the-bleachers buddy? Fellow Scotch-tasting groupie? Yes, as it turned out, someone else in my life was an apple crackerjack.

"I love working the apple festival," Daphne practically gushed when I asked her about it. Daphne is really not a "gusher," so once again I was surprised by the passion that ferments inside apple people. "People who are into apples are so romantic about it," she said. "They get dreamy because they associate apples with memories and stories."

Their reactions to the apples are emotional, she added, like "Oh, my grandmother used to have one of those apple trees near her back porch and I remember sitting under it when I was a kid."

Often, it's the scent of the apple that triggers memories, and Daphne recalled one woman who sought to identify an apple this way, holding it in one hand, smelling it, and comparing it to each apple on the display table. But what struck Daphne the most were the stories and ultimately the nostalgia that apples—especially heritage apples—evoked. I'd also come across this passion as I talked to apple people and read articles they'd written. For example,

OPPOSITE:
A collection of Gravenstein apples at Apple Luscious Organic Orchard shows the diversity of shapes and sizes of just one variety of apple. The shapes have too much variation for commercial growers, but as "apple man" Harry Burton points out, the apples are delicious! Gravenstein is an all-time favourite for many heritage apple lovers.

I found an article, "Singing the Praises of Gravenstein," by Dr. Jim Rahe, a Simon Fraser University professor of plant pathology.

"Who hasn't heard of . . . heritage variety Gravenstein?" he enthused. "The standard for crisp and juicy—Gravenstein! Sprightly flavour—Gravenstein! Great for sauce, for pies, for juicing, for drying—Gravenstein! Apples like Grandma used to bake—Gravenstein!"

He went on to write an eighteen-hundred-word article—part history, part anecdote, but mostly informative—discussing and comparing test plantings he did in the 1980s of twelve different strains of Gravenstein. He concluded by saying, "I relate this story to encourage our members to compare strains of old heritage vari-eties. Who knows, you just might find a nugget among the stones. And talking about them at parties can be a lot of fun!" [4]

Although the article itself was interesting, it was the tone that captivated me; again, the unbridled passion that heritage apples inspire. I think it's this, more than anything definitive, that describes the "what" of heritage apples. On a purely subjective level, heritage varieties are those apples that link us to our past; apples that elicit memories and history and passion. For those who want a more conclusive definition of heritage apples, that issue remains under debate. One school of thought says the "cultivar" (defined as "a variety of plant that originated and persisted under cultivation") must be over one hundred years old, other people say fifty. Some gardeners consider 1951 to be the latest year a plant can have originated and still be called heritage, since that

year marked the widespread introduction of purposeful controlled hybridization.

"Really, a heritage variety is one that's been around for a long time; it has a heritage that is part of our heritage," said apple identification expert Clay Whitney. "You can't really put a date on it."

Dr. Bob Weeden, a passionate heritage apple grower and former biology professor, agreed. "No one, I think, can 'know' when heritage varieties gave way to modern ones, as it is definitional. I like a definition that notes the distinction between, on one hand, varieties that resulted from natural chance and were found along hedgerows or in seed orchards where folks grew thousands of chance hybrids in hopes of finding a good one to propagate, and [on the other hand] those coming from purposeful crosses of untampered pollen. There won't be 'a date' for that, as the two techniques overlapped—still do!"

To understand all this, one needs to know a bit about apple sex, which in the wild gives birth to seedlings that create entirely new strains of apples. Apple sex in controlled environments by apple breeders also results in new varieties via seedlings, but those are a little more predictable. And, of course, apple varieties can be reproduced exactly and deliberately via grafting. Apple reproduction is an example of something called "heterozygosity"—rather than inheriting DNA from their parents and creating apples with the same characteristics, apples are instead different from their parents, sometimes radically.

The good news for the apple, as Michael Pollan notes in *The*

Botany of Desire: A Plant's-Eye View of the World, is that this variability has allowed it grow all over the world under extremely varied climates and conditions.

"Wherever the apple tree goes," Pollan writes, "its offspring propose so many different variations on what it means to be an apple—at least five per apple, several thousand per tree—that a couple of these novelties are almost always bound to have whatever qualities it takes to adapt and prosper in a new home." [5]

In earlier times, apples were often grown from seeds, which can be carried more easily than seedlings over distances. Pioneers brought them to North America, sometimes saving the seeds from apples they consumed during their voyage across the ocean. Today, most new apple cultivars originate as seedlings, which either arise by chance or are bred deliberately by crossing cultivars with promising characteristics. The words "seedling," "pippin," and "kernel," in the name of an apple suggest it originated as a seedling.

But how do apple breeders create new apples?

"How do apples have sex?" I wondered aloud.

"Soft-light and moody music?" Bruce offered. "Probably not candles, though."

He is amusing at times.

This is how it happens. Whether or not an apple is reproducing in the wild or assisted by a breeder, the seed is a cross between the female parent (the flower of the tree from which the apple came) and the male parent (the variety of apple tree that produced the pollen). So in the wild, a Cox's Orange Pippin could be pollinated by a nearby King, and the result would be a cross between the two. It's hard, therefore, to determine the parentage of a seedling found growing at the side of the road because the origin of the parent blossoms is unknown. Apple breeders use the same principle, but much more deliberately. Here, pollination is controlled by covering the apple blossoms with fabric to prevent anything random occurring. When the blossoms are open and ready for pollination, the breeder uncovers them and manually applies pollen from the blossoms of a known variety before covering them up again. The seeds from these hand-pollinated apples are collected and planted. Hundreds or thousands of seeds can be planted in the hopes of finding a desirable variety (grower Harry Burton says the odds of a seed producing a better variety are less than one in ten thousand) and it takes four to eight years for each of those trees, started from seed, to produce fruit. So apple breeding is a long-term, large-scale project.

Once you have an apple that you want to reproduce, you turn to grafting, a technique in use for thousands of years, most likely discovered by the Chinese in at least 2000 BC. Grafting, like cloning, ensures that the new tree is exactly like the parent tree. A detached shoot or twig from the desired tree, called the scion (pronounced *SY-uhn*) is attached to the rootstock, which is the working part that interacts with the soil to take root and nourish the new tree.

Once when I visited Harry, I had a chance to watch him deftly graft about a dozen apple trees. The room, in an outbuilding near his house, was chockablock with floor-to-ceiling boxes, obviously

used for apple distribution come harvest. But Harry had carved himself out a small space, where he sat, pulled out his knife and calipers, and showed me how to make the perfectly angled cuts to create the tongues that would allow the scion and rootstock to join and eventually grow together. The trick, Harry explained, is to get enough of the cambium (the growing layers, just under the bark) touching, so that they can meld together and allow fluids to pass up and down the tree across the graft. Once the scion and rootstock are joined, grafting tape strengthens the join and keeps in moisture. The new seedling is put in water and—the sooner the better—planted. Harry plants grafted seedlings in April, the tape

When compared to a hand, the size of a Wolf River apple becomes apparent. No wonder they are a favourite for apple pies.

Bright red apples shine on a dwarf tree at the UBC Botanical Garden in Vancouver. Every year, thousands of people flock to the annual apple festival held there.

comes off in July, and in a few years the two parts will have grown together, producing a single tree.

Grafting can also be accomplished in August using a T-bud graft. Here, a mature bud at the base of a leaf stem from the desired apple species is inserted below the bark of the rootstock in a T-shaped cut about fifteen centimetres above the ground. The entire stem is wrapped several times with grafting tape to seal the bud in place, and keep it moist and tight against the rootstock cambium layer. This dormant bud will not grow until the following spring, when the tree is cut off above the bud, and the new bud starts to grow.

Rootstock is chosen to control the final size of the tree, and over the years, the effect of different rootstock on the growth of scions has been thoroughly studied. Original old apple trees tower in height. But this makes it difficult to harvest the apples, so many people choose to grow apple trees on medium rootstock or, even more common, on dwarfing rootstock. For commercial growers— or hobby growers—the dwarfing rootstock results in small trees that produce a high volume of fruit. The trees are easier to prune and pick, and the number planted per acre is much higher. However, many growers—like Harry and Bob—don't like dwarfing rootstock because the reduced vigour of these trees results in the need for more attention in terms of watering, staking, and fertilization.

Most apples are bred for eating fresh (dessert apples), though some are cultivated specifically for cooking or producing cider. Cider apples are often too tart and astringent to eat fresh, but they give the beverage a structure of flavour that dessert apples don't.

Over the years, the types of popular apple cultivars have changed to meet the demands and tastes of consumers (which vary among different regions in the world) and the needs of commercial growers. Consumers generally favour apples that are sweet and crisp, while growers want varieties that yield high numbers, ship and store well, and are resistant to disease. They want apples that are colourful, pretty, and well-shaped. Heritage apples can be oddly shaped, russeted, and have a variety of textures and colours. Although some may taste better than modern cultivars, they may be harder to grow or lack good storage or transportation capabilities.

A few heritage cultivars are still produced on a large scale (like Cox's Orange Pippin in the United Kingdom and, of course, McIntosh in North America), and many have been kept alive by home gardeners and people like Harry and Bob who sell directly to local markets. In grocery stores today, there's a limited range of commercial apple varieties available, especially compared to the nineteenth and early twentieth centuries, when the diversity was huge. But there are many groups all over the world working in different ways to preserve heritage varieties. One of these, a group called LifeCycles in Victoria, BC, runs the Fruit Tree Project, which links people who have surplus fruit trees in their yards with volunteers who are able to harvest them. The harvested produce is split among homeowners, volunteers, food banks, community groups, and businesses.

According to its website, the Fruit Tree Project is a "celebration of a simpler time when we fed ourselves from our own orchards and gardens, timed the passage of seasons by the ripening of fruit and discussed pie recipes over the fence with our neighbours."

It also points out some of the "how" and "why" heritage apples have fallen into disfavour and disuse in modern society:

"In the 1800s, Victoria became the legendary fruit-growing centre of BC. The stately old fruit trees in backyards across the city are the legacy of orchards that once flourished in the region's mild climate and fertile soils. The sad part of the story is that the apple has seen its heyday. With our busier lifestyles, few of us find time to cultivate and harvest apple trees and many of Victoria's old apple trees now drop their annual loads on someone's lawn. While the wasps gorge on homegrown fruit, Victorians bring home bags of shiny Granny Smiths shipped from the Okanagan, the US or New Zealand. Such market varieties are chosen for durability and good looks rather than flavour, quality or historical importance. As a result, many varieties that were valued in the past for exceptional texture, taste, good storage ability or tradition are being lost. With them goes centuries of history and careful selection along with our ability to discern and appreciate subtle nuances of their flavours." [6]

There are other reasons why people want to preserve heritage apples (and other heritage fruit). Some are motivated to grow heritage varieties because they taste better and may even have higher nutritional value—such as Harry's red-fleshed apples. Biodiversity is a big factor for many, such as Clay Whitney who sees genetic diversity as crucial in fighting disease like canker. Others grow heritage apples for more cultural reasons, like the preservation of living history, while some just want the challenge of growing rare varieties or have a particular interest in traditional growing techniques.

And then, of course, there is that passion that heritage apples inspire—that enthusiasm exhibited by Daphne and those "dreamy" apple people she meets year after year at the festival. Like the exuberant writer Jim Rahe, Daphne cited Gravenstein as her favourite apple—the red variety in particular. In the heritage apple orchard of her dreams, said Daphne (who is a landscaper by trade), she would grow a tree each of Red Gravenstein, King (for juice and applesauce), Granny Smith (because it keeps for a long time), Northern Spy ("for fun"), Pippin ("for candy"), and Belle de Boskoop (for cooking).

Daphne's choice apples include some of those listed by Harry, who polled numerous apple growers to determine which are the biggest heritage apple sellers. He names the top five as Gravenstein, Cox's Orange Pippin, King, Goldengelb, and Holstein.

Even without her dream orchard, Daphne will continue to enjoy the fruits of her labour at the apple festival and, apparently, she and I can add "apple talk" to our sports-watching days and Scotch-tasting nights. Maybe we can even eat a Red Gravenstein or two.

BELLE DE BOSKOOP
HOLLAND, 1856

PHOTO: DERRICK LUNDY

Taste and appearance: Medium to large fruit with skin that is gold and red-striped, often with a lot of russeting. The flesh is pale green, juicy, sharp, rich-flavoured, and very aromatic.

Use: Eating fresh and cooking. Keeps its shape when cooked. Good for pies.

History: This dual-purpose Dutch apple was discovered at Boskoop, Holland, in 1856 and is thought to be a mutation of Reinette de Montfort.

Growing and harvesting: The tree can be slow to bear fruit. It's picked in late September and early October, and keeps six months.

Other: Particularly high in vitamin C.

GRANNY SMITH
AUSTRALIA, 1860s

Taste and appearance: A smooth, green apple with a sharp-but-sweet flavour.

Use: Eating fresh and cooking. Keeps its shape when cooked and works well in salads with its crisp, slightly acidic flavour.

History: In the Sydney suburb of Ryde, Australia, Granny (Maria) Smith is said to have thrown some rotting French crabapples onto a creek bank near her orchard, now called Granny Smith Memorial Park. From those apples a chance seedling produced the green apple, which, by 1868, was known locally as good for eating and cooking.

Growing and harvesting: The fruit matures late so needs a warm climate to ripen. It's picked in late October and best eaten after January. It keeps until April.

Other: The tree is susceptible to scab and mildew. By the 1960s Granny Smith was practically synonymous with "apple" and the variety was used by the Beatles as the logo for their company Apple Records.

GRAVENSTEIN
DENMARK, 1669

PHOTO: CLAY WHITNEY

Taste and appearance: Red or green, fairly large, roundish or irregular, and prominently ribbed. The flesh is cream-coloured, crisp, juicy, melting, and aromatic with old-fashioned tart and sweet flavours.

Use: Eating and cooking. Described as "unexcelled" for applesauce and pies, it keeps its shape while cooking.

History: Thought to have been growing in Denmark (or possibly Northern Germany) in about 1669, it was first grown in California in 1820 and in Nova Scotia in 1835.

Growing and harvesting: In North America, it has a reputation for being fussy, and is said to do better in areas where the climate is closer to the milder winters and cooler summers of northern Europe. Picked in late August, it keeps for two months. Can still be used in cooking after three months if kept refrigerated.

Other: The apples ripen over a period of weeks and drop readily, a boon for the home gardener, but a commercial drawback. Gravenstein was declared the national apple of Denmark in 2005.

PHOTO: CLAY WHITNEY

HOLSTEIN
GERMANY, 1918

Taste and appearance: Fairly soft, sweet-tasting, with a slight pineapple flavour. Red-orange flush over a yellow skin and some russeting.

Use: Good for eating and cooking (especially pies) and noted for its excellent orange-yellow juice.

History: Discovered by chance in Germany in 1918, in the Holstein region, possibly from Cox's Orange Pippin. Also known as a Holstein Cox.

Growing and harvesting: Described as vigorous and easy to grow, a great keeper, and able to retain its qualities for a full five months after harvest if kept chilled.

Slightly Curried Apple-Squash Soup

4 big King apples

1 medium-to-large acorn squash

vegetable broth, or water, to cover apples and
squash

1 onion

4 garlic cloves

2 Tbsp red curry paste, or 1 Tbsp curry powder

2 Tbsp cumin

salt and pepper, to taste

Peel and core or seed apples and squash. Cut them up, cover with broth or water, and simmer until soft and dissolving. In a separate pan, chop onion and garlic and sauté until translucent. Add spices. Add onion mixture to apple/squash mixture, and mix with a hand blender until smooth. If it's too thick, add some more broth until it's just right.

Recipe provided by Daphne Taylor, Salt Spring landscaper and caterer

Applesauce

12 Gravenstein apples, quartered

½ cup sugar

2 Tbsp unsalted butter

1 tsp cardamom

1 star anise

1 tsp lemon juice

2 tsp cinnamon

½ cup water

Place all ingredients in a large pot; bring to a boil over high heat. Reduce to a simmer and cook, covered, for 10 minutes. Add more liquid if needed. Cook for an additional 20 minutes, until apples are very soft. Remove star anise. Transfer to a food processor and purée until smooth. Serve hot or cold.

Recipe provided by Steve Glavicich, chef/owner, Braizen Food Truck, Calgary

CHAPTER THREE

From the Garden of Eden

A basket of
brightly coloured,
delicious-looking
Newton Pippins.
PHOTO: DERRICK LUNDY

A 150-year-old Lemon Pippin tree still produces fruit at Woodside Farm in Sooke, BC. In the background is one of two remaining houses built in 1884 by John and Ann Muir, who obtained land from Sooke pioneer Captain Grant when he returned to Britain in 1853.

Apple Facts

Some "great moments in apple history" compiled by Mitch Lynd:[1]

6500 BC: Remains of apples found among excavations at Jericho in the Jordan Valley are dated to this time period.

5000 BC: Feng Li, a Chinese diplomat, gives up his position when he becomes consumed by grafting peaches, almonds, persimmons, pears, and apples as a commercial venture, according to *The Precious Book of Enrichment*.[2] Agriculturalists are charmed; naturalists are alarmed.

1500 BC: A tablet found in northern Mesopotamia records the sale of an apple orchard by Tupkitilla, an Assyrian from Nuzi, for the significant sum of three prized sheep.

1665 AD: Sir Isaac Newton watches an apple fall to the ground and, wondering why it falls in a straight line, is inspired to discover the laws of gravitation and motion.

I met apple history head on as I meandered through a 140-year-old orchard on a sunny afternoon in late September. The thick-trunked trees soared above me, unencumbered by the dwarfing techniques of modern apple growing, some of them completely hollow inside and yet still producing apples, as their life force is contained in the cambium layer inside the bark. I thought of the history that had travelled past these trees, from pioneers clearing the land, building shacks, and carving out their simple livelihoods, to the people of today, driving by in sleek cars, with smartphones and GPS and tablet computers. I imagined the progression of dialogue that had swirled under these limbs, the changing and not-so-changing worries and wonders expressed. (For example, I picture lovers from different eras parting beneath the canopy, one whispering, "Goodbye, shall we meet here next week?," the other saying, "Later. Text me.")

But then my mind's eye went even further back, trying to imagine how the wild apple orchards of Kazakhstan must appear. There, with no influence of cultivation, the apple trees apparently grow to sixty feet tall and host a sweeping array of apple shapes, ranging in size from marbles to softballs, and in colour from yellow, to green to red and even purple. The vista must be otherworldly, like something from the pen of Lewis Carroll.

As a lover of the wild apple, Henry David Thoreau would have appreciated it. In an 1862 essay commissioned by *The Atlantic* and called "Wild Apples," he enthused, "I love better to go through the old orchards of ungrafted apple-trees, at whatever season of the year—irregularly planted: sometimes two trees standing close together; and the rows so devious that you would think that they not only had grown while the owner was sleeping, but had been set out by him in a somnambulic state."[3]

Russian botanist Nicolai Vavilov is credited in the early 1930s with identifying the birthplace of the apple as Alma-Aty (or Almaty), Kazakhstan, in eastern Asia, near the western border of China. (However, translated into English, Alma-Aty means "father of the apple," so someone earlier must have had an inkling of what was going on.) Vavilov's research sank amid politics in the Soviet Union and it wasn't until decades later that his work was extended by scientists such as Dr. Barrie Juniper and others, from the Plant Sciences Department of Oxford University.

Juniper, who co-authored *The Story of the Apple*[4] with his student David J. Mabberley, made several trips in the 1990s to the huge wild-fruit forests on the mountain slopes near Alma-Aty, as well as in neighbouring Uzbekistan and other areas in eastern Asia. The Oxford University research occurred alongside work undertaken by a team of Cornell University scientists, who collected specimens from Uzbekistan, Kazakhstan, Kyrgyzstan, and Tajikistan for the Plant Genetic Resources Unit at Geneva, New York (the site of the largest apple, grape, and cherry repository in the world).

DNA studies confirmed that the cultivated apple had its origins in Kazakhstan. As summarized in Juniper's article "The Mysterious Origin of the Sweet Apple,"[5] the apples in this isolated area—surrounded by mountains and deserts—evolved by natural selection from tiny fruit eaten and distributed by birds and small animals into bigger, sweeter fruits attractive to larger wild mammals such as pigs, bears, and horses. These animals spread the seeds throughout the Middle East along animal migration tracks, which later became trade routes known as Silk Roads.

Apples were further spread along these routes by human travellers, who picked the largest and tastiest of the wild-growing apples to carry west with them, dropping seeds along the way. Wild seedlings would have grown and hybridized with other species such as the European crab, eventually producing apples throughout Asia and Europe.

But true cultivation didn't occur until the invention of grafting, which allowed the Romans and Greeks to select and propagate the best varieties. By the first century AD, the Romans had cultivated twenty-three types of apples, some of which they took to England. The Lady apple is thought to be one of these.

After apple-growing came to England and France via the Romans, it spread to North America with the colonists in the seventeenth century. The first apple orchard in America was said to be near Boston in 1625, the first US commercial trade began in the 1740s with exports to the West Indies, and by 1871 the fame of the Newton Pippin, an apple discovered in a Flushing, New York, cider orchard, had already spread to Europe.

The earliest immigrants to America brought grafted old trees with them, but many of these did not thrive in the North American climate. The colonists also planted seeds saved from apples eaten

Mike and Marjorie Lane have been managing Ruckle Farm since 1990. They aim to run it in the same way as the homesteading Ruckle family did, raising sheep and lambs, cows, hens, and turkeys, and growing hay. They also harvest and sell numerous varieties of heritage apples.
PHOTO: DERRICK LUNDY

during their Atlantic voyage and some of those seedlings—called pippins—eventually prospered.

One the most colourful figures emerging from this time was the larger-than-life John Chapman (AKA Johnny Appleseed), credited with planting over ten thousand square miles of trees between Pennsylvania and Fort Wayne, Indiana, where he died in 1845. Stories over the years have elevated Chapman to legendary status, claiming he travelled barefoot, wore a saucepan as a hat, preached the word of the Swedenborgian Church, cared deeply about animals (including insects), and was, in fact, kind and generous to all.

The best reading, when it comes to Chapman, is from Michael Pollan, who, in *The Botany of Desire*, sets out on Chapman's trail, determined to find the truth amid the "Disneyfication" of the legend. Today, Pollan says, the apples and the man have suffered a similar fate in the years since they travelled down the Ohio River together in a double-hulled canoe. "Both then had the tang of strangeness about them, and both have long since been sweetened beyond recognition . . . Chapman transformed into a benign Saint Francis of the American frontier, the apple into a blemish-free, plastic red saccharine orb." [6]

Chapman did not spread apple seeds randomly everywhere he went. Instead, he deliberately moved west, keeping ahead of settlers, and planting nurseries in wilderness areas he believed suitable for settlement. When the settlers arrived, he had trees ready to sell. In some areas, settlers were required by law to plant orchards as a condition of their land deeds, the goal being to prevent speculation by encouraging homesteaders to stay in one place. Since a standard apple tree could take ten years to fruit, an orchard was seen as a mark of permanency.

Popular mythology around Johnny Appleseed also skirts the "point" of all these apples. As Pollan notes, apples grown from seed are rarely sweet or tasty; people who wanted edible apples grafted trees of their choice. Chapman's seedlings mostly produced sour apples, used almost exclusively for cider, the most common beverage of the time, even for children.

"Apples were something people drank," Pollan says. "Johnny Appleseed was actually bringing the gift of alcohol to the frontier." [7]

Cider could be made by anyone with a press and a barrel, and even Puritans were able to give it a "theological free pass," notes Pollan, since the Bible's Old Testament warns against the temptation of grapes, but says nothing about the apple or the strong alcoholic drinks that can be made from them. "America's inclination towards cider is the only way to explain Chapman's success." [8]

In addition to cider making, apples were an excellent commodity for early settlers since they stored well and had many uses. They could be made into apple butter and apple pies, or cored and dried for even longer storage.

So apples moved west with the settlers. In 1824, Captain Aemilius Simmons brought seeds to Fort Vancouver in Washington: now, with the advent of irrigation technology in the twentieth century, Washington is the top-producing apple state in the US.

In Canada, apples arrived with the French, early in the seventeenth century. Apples may have been grown at Port Royal in Nova Scotia as early as 1606, and it's certain that varieties were growing near Annapolis Royal by the 1630s. Early records also show that apple trees existed in LaHave, Acadia, by 1635. In *The Apple: A History of Canada's Perfect Fruit*,[9] author Carol Martin says Samuel Champlain brought young saplings for planting when he landed in 1608 at the "point of Quebec" and constructed the first buildings of what later became Quebec City. Gardening was one of Champlain's passions.

However, during this time apples were grown mostly by the Acadians—a 1698 census counted 1,584 apple trees at Port Royal alone. Like their American counterparts, most of these early apple growers used the bulk of their fruit for cider.

Canada's first known cultivar, Fameuse (also known as Snow), has been grown in Québec for centuries and is possibly the parent of McIntosh. It arose from seed or possibly a young seedling brought from Normandy, France, during this time. From the mid-1700s to the 1850s, Fameuse was the most common apple tree grown in Canada, and the fruit was exported to England in large quantities. For some unknown reason, a massive destruction of Fameuse apple trees occurred in Quebec orchards in the 1860s, reports Canadian apple enthusiast Eric Rivardon on the comprehensive heritage apple website orangepippin.com, and new varieties were introduced, such as Wealthy, Baldwin, and some species from Russia.

According to Carol Martin, apples first arrived in the west of Canada via seeds tucked into the vest pocket of Governor George

Crabapples are the only variety of apples native to North America. These are Dolgo apples, grown in Edmonton. According to Gabor Botar, they are the "best crabapple for jelly."
PHOTO COURTESY GABOR BOTAR

Simpson, who oversaw the merging of the Hudson's Bay Company and the Northwest Company in 1821.

"Why were apples so desirable? For early immigrants, they provided a sweet and healthy addition to their sometimes meagre diet, and dried they were available for use all year round. But, most of all, apples could easily be turned into cider, the ubiquitous drink of the times," says Martin. "Native people too welcomed these cultivated apples that were such an improvement over the small, hard crabapples they were accustomed to." [10]

Currently in Canada, the main commercial apple-growing regions are in Nova Scotia, southern Québec, Ontario—near Lakes Ontario and Erie—and the southern valleys of BC. However, all provinces except Newfoundland have some commercial production.

The Pacific crabapple is the only apple native to BC. It grows quite widely, particularly near water, and produces small green fruits.

Historically, these crabapples were eaten by First Nations, and old trees have been found in some abandoned Haida villages. Historians are unsure if they were intentionally planted there or grew from discarded pips.

The first places in BC to cultivate apples in the 1860s were Salt Spring Island, other nearby Gulf Islands, and areas around Victoria on Vancouver Island. According to archival research undertaken by Harry Burton, Salt Spring was the major fruit-producing area in BC, and by 1895, there were roughly 450 residents with about 4,600 fruit trees; this didn't include smaller orchards with under 200 trees.

In 1895, pioneer Samuel Beddis planted an orchard of 500 trees using rootstock created by planting apple seeds saved from apples eaten on the trip from England. According to Harry's website, he "later he grafted these young trees with scions from 40 or so varieties shipped from Ireland. Each scion had been sent safely traveling by mail, embedded in an Irish potato. His and other orchards flourished in the mild lush climate."[11] A partially burnt notebook that belonged to Samuel Beddis includes carefully drawn diagrams of his orchard and a list of the fruit trees. The list is incomplete and misspelled, but gives an idea of some old varieties he grew, including Canada Reinette, Baldwin, Blenheim Orange, Wealthy, Duchess of Oldenburg, and Gravenstein.

Other large orchards of between 250 and 1,600 trees were also planted at the time, with farmers claiming growing conditions were better than in Britain, due to warm summers and mild winters.

Apples sold for two cents per pound or seventy-five cents per bushel.

"Some of the Canada Reinette [trees] were so loaded with fruit," Harry said, "they had to have every branch propped . . . yields per tree were twenty-four boxes of fifty pounds each."

An article entitled "The Islands—A Well Known Fruit District," published in 1908 in Victoria's *Daily Colonist* newspaper said, "Some of the finest fruit grown in the region of Vancouver Island comes from the islands of Salt Spring, Pender, Main [*sic*], and Galiano. These islands are extremely well adapted for fruit cultivation, and agriculture and fruit growing have long been their staple industries." It notes that one of the drawbacks to the success of fruit-growing was the "tendency to attempt too great a variety," but this had "more or less been corrected" with the conclusion that King, Baldwin, Northern Spy, Jonathan, and Gravenstein "are the varieties of apples best suited to the islands." [12]

The article also mentions the challenges related to transporting apples—"transportation facilities being one setback in the past, and even yet a serious detriment." Indeed, apples were exported from Salt Spring via rowboat, five miles across the water to Sidney.

About 1895, one of the largest and most productive farms on Salt Spring belonged to Henry Ruckle, whose descendants donated the non-farm portion of the land to the province of BC for the creation of Ruckle Park in 1977. It's now the oldest continuously operating family farm in BC. Current farm manager Mike Lane and his wife, Marjorie Lane, have been involved with Ruckle Farm since 1990, and each year showcase their heritage apple varieties at the island's apple festival. Many of the trees still producing apples are well over one hundred years old, and it was this orchard that I wandered through, thinking of all the events that have occurred under those apple tree limbs.

One of Salt Spring's beloved old-timers, Lotus Ruckle, who died in 2009, married Henry Ruckle's son, Gordon, in 1930, moving one kilometre down the road to Ruckle Farm from her stepfather's farm, which today grows heritage apples under the banner of Wave Hill Farm. Harry Burton knew Lotus well, and she told him of a Blue Pearmain tree in her childhood orchard, which had apples so delicious that the family would not sell any, not even to their neighbours. She said her stepdad, island pioneer Cory Menhinick, made "cider equal to the best champagne from his prized Gravenstein apples." Cider sold for seventy-five cents a gallon in the 1920s, and that was the farm's main income, said Harry. Lotus' job was to feed the apples into the grinder.

Agriculture, including fruit production, continued to spread throughout BC, spurred by gold rushes, railway production camps, and the mining industry. By the 1880s, the Okanagan Valley had developed a specialized fruit industry, but it wasn't until the 1920s that its apple production began to surpass Salt Spring's. This was due to improved irrigation, rail transportation, and rural electricity in combination with the area's hot summer temperatures, good soil, and cheap land.

Bob Weeden has a "fascinating little book," published in 1912, by J.T. Bealby entitled *How to Make an Orchard in British Columbia*,

Award-winning apples at the annual Salt Spring Fall Fair, which is held on the island every September.
PHOTO: DERRICK LUNDY

in which an advertisement offered "thousands of acres of wonderfully productive land for sale in southern British Columbia at extremely low prices." The cost was "62.5 cents per acre cash and 62.5 cents per acre for seven years thereafter."

Okanagan apple-growing history credits a priest named Father Pandosy with planting the area's first apples in 1859 at his Catholic mission in Kelowna. His apples were for use at the mission only, however, and the Okanagan's first commercial apple enterprise didn't appear until several decades later. This was started in the 1890s by Lord and Lady Aberdeen, who bought a thirteen thousand-acre ranch in Coldstream, near Vernon, plus other huge tracts of land. They planted one hundred acres of apple trees at each location.

"The Aberdeens spent a considerable amount of money and time encouraging others to start fruit farming in the Okanagan Valley," noted one document. "Lord Aberdeen was so convinced of the profitability of apple growing that he later subdivided some of his Coldstream Ranch into ten to forty acre parcels to be sold for commercial orchards." [13]

But farming in the Okanagan was difficult, as farmers dealt with insects, winter freezes, and poor transportation. And often, orchardists planted apple varieties unsuitable for the area. Irrigation was an even bigger problem, with a lack of adequate rainfall in many areas, and no government projects providing water. This all changed with the damming of the Columbia River; dams built in the 1930s brought cheap water and irrigation to the big Okanagan orchards, which effectively took over the market.

Several factors over the past 150 years have lessened the apple's popularity. Once North American pioneers had access to cane sugar from the West Indies, the apple was no longer the sweetest item on the menu. Cane sugar could also be used in alcohol production, again reducing the need for apples. In more recent times, the huge acreages of citrus in Florida and California, with fruit sent north by truck and rail, have brought fatal winter competition to people who invested in cold storage, for provision of apples throughout the winter. Cheap oil enables the importation of papayas and mangoes and bananas from far away at prices apple-country people, paying high wages for labour, can't match except with big, machine-driven, efficient industrial orchards.

Still, the future for apples could yet be bright, as rising transportation costs make buying locally produced fruit more affordable. And of course, movements like "Slow Food" and "Eat Local" can only help raise apples, and especially heritage apples, back to their former popularity.

LADY
FRANCE, 1628

PHOTO: CLAY WHITNEY

Taste and appearance: Very small, with a pale green skin blushed with layers of red and nearly invisible white freckles. Though it's firm to the touch, the flesh is tender, not crisp. The flavour is sweet-tart and subtle. It's highly aromatic.

Use: Eating fresh, decorating, and cooking. It is best cooked (baked, caramelized, or roasted) to bring out the flavour. Also used decoratively, especially at Christmas and often in wreaths.

History: Thought to be the oldest recognized variety under the name Pomme d'Api. It was originally documented in early Rome during Etruscan rule (approximately 700 BC). The first reference to it as a Lady apple was in 1628 during the French Renaissance.

Growing and Harvesting: Available September through January. Left out, it dries nicely; refrigerated, it lasts up to four weeks.

Other: Considered a cheerful holiday fruit and fun to eat (two bites is all it takes). Don't peel Lady apples because the peel adds to the winey, semi-sweet taste of the flesh.

Taste and appearance: Yellow, green, and semi-russeted, with a sweet, tasty flavour. The extent of russeting is usually fairly light; the underlying light yellow or green skin is readily visible and may be flushed red.

Use: Eating fresh, but primarily cooking.

History: Despite the name, Reinette du Canada is an old apple from France.

Growing and Harvesting: Ripens in mid- to late season and stores well for one or two months.

Other: It is one of the most widely grown russet varieties in France, readily available in supermarkets and village markets simply as "Canada," although this term is also used for the more russeted Reinette Grise du Canada.

FAMEUSE
QUEBEC, EARLY
1800s

Taste and appearance: A small, red apple with brilliant white flesh and a distinct sweet flavour.

Use: Eating fresh, juice, and cider.

History: It originates in Quebec, where it was grown commercially on a large scale in the 1800s.

Growing and Harvesting: Ripens late in the season and keeps well for one to two months. Able to tolerate extreme winter cold, it has good disease resistance with the exception of being highly susceptible to scab.

Other: Probably the ancestor of McIntosh, which in turn has led to many other varieties, all characterized by crimson skin colours, sweet white flesh, and a unique, sweet flavour.

PHOTO: DERRICK LUNDY

NEWTON PIPPIN
NEWTON, LONG
ISLAND, NEW YORK,
MID-1700s

Taste and appearance: Medium-sized green apple with cream-coloured flesh.

Use: Eating fresh, cooking, juice, and cider.

History: One of the oldest American varieties, it was well known in the eighteenth century, probably raised as a seedling by early settlers on Long Island. It was introduced from the US to England in the 1750s.

Growing and Harvesting: A winter apple, it's picked in late October but tastes better after being stored for one to two months. The tree can take several years to start bearing apples, even on dwarfing rootstock, and it is susceptible to most of the usual apple diseases.

Other: Newtown Pippin was popularized by people such as Thomas Jefferson and George Washington.

Maple Apple Butter

2 Tbsp unsalted butter

3 lb assorted apples, peeled, cored, and
    quartered

2 cups apple cider

1 tsp lemon zest

juice of ½ a lemon

1 cup maple syrup

½ tsp ground cinnamon

In a Dutch oven over medium heat, melt butter and add apples. Cook apples until slightly softened, about 5 minutes. Add cider, bring to a boil, and reduce heat to a simmer. Let simmer, partially covered and stirring occasionally, until soft, about 30 minutes. Preheat oven to 250 degrees. Mash softened apples with a potato masher. Remove from heat and add lemon zest, lemon juice, maple syrup, and cinnamon. Using a hand blender, purée mixture until smooth. Pour apple mixture evenly into a baking dish. Transfer to oven and bake, stirring occasionally, until thickened and reduced, about 3 to 3½ hours. Let cool before serving. Apple butter may be kept, refrigerated in an airtight container for up to 5 days.

Recipe provided by Steve Glavicich, chef/owner, Braizen Food Truck, Calgary

Chocolate-Covered Lady Apples

8 oz semi-sweet chocolate chips

12 Lady apples, washed and completely dried

½ cup chopped pistachios or shredded coconut

Line the counter top with parchment paper. Place the chocolate chips in a microwave-safe bowl and melt them on low power for 1 minute. Stir and place them back in the microwave for 30 seconds at a time until melted. Hold the apples by the stem and dip them three-quarters of the way into the chocolate. Lift and twirl them lightly to get rid of excess chocolate. Sprinkle with pistachios or coconut and set on parchment paper. Allow the chocolate to dry for 30 minutes before consuming or wrapping.

    Note: When buying apples, be sure to pick ones with stems. They look prettier and are easier to dip. Also, if you wish, you can flavour the melted chocolate with your favourite liqueur before dipping.

Recipe by *Dinners and Dreams* blogger Nisrine Merzouki (www.dinnersanddreams.net)

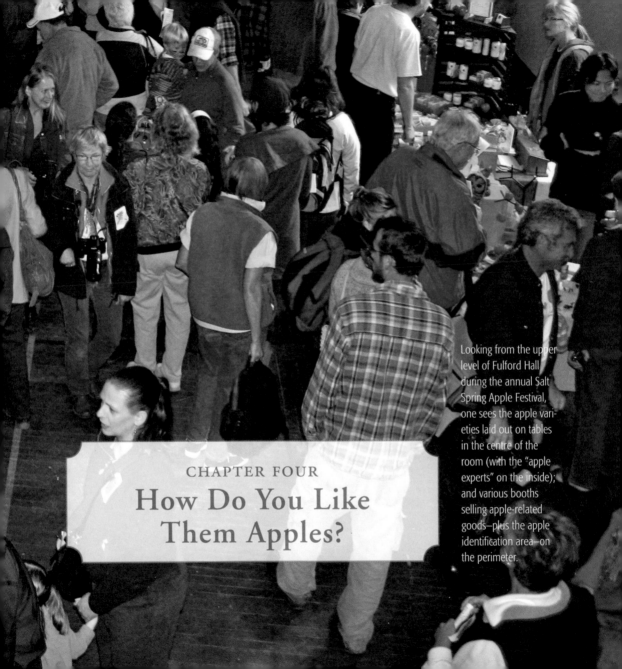

CHAPTER FOUR
How Do You Like Them Apples?

Looking from the upper level of Fulford Hall during the annual Salt Spring Apple Festival, one sees the apple varieties laid out on tables in the centre of the room (with the "apple experts" on the inside); and various booths selling apple-related goods—plus the apple identification area—on the perimeter.

Apple Facts

Facts gathered by the University of Illinois:[1]

Two pounds of apples make one nine-inch pie.

Apple varieties range in size from a little larger than a cherry to as large as a grapefruit.

Most apples can be grown farther north than most other fruits, because they blossom late in spring, minimizing frost damage.

The world's largest apple peel was created by Kathy Wafler Madison on October 16, 1976, in Rochester, NY. It was 172 feet, 4 inches long. (She was 16 years old at the time and grew up to be a sales manager for an apple tree nursery.) But common advice says don't peel your apple. Two-thirds of the fibre and lots of antioxidants are found in the peel.

In colonial times, apples were sometimes called "winter banana" or "melt-in-the-mouth."

It was hard to find parking at Fulford Hall where the doors to Salt Spring Island's annual apple festival were just opening. A crush of people lined up at the ticket centre, waiting to pay a small festival fee in exchange for a big apple sticker to plaster on their chests, as well as a map of participating island farms and a page of select apple facts from Captain Apple himself. That would be Harry Burton, organizer of this event which each year draws over twelve hundred people, making it one of the biggest annual events on little Salt Spring Island. Harry was hovering by the ticket booth, recognizable by the big red apple painted on his face.

Since I'd dropped by the hall the previous night to watch the set-up, I wasn't entirely surprised by the sight inside, which was highlighted by two massive lines of tables—absolutely chockablock with apples—right down the full length of the hall. About 50 volunteers, some of whom had been doing this for a decade or more, came, ate dinner, talked apple talk, and then got down to work, setting out 302 varieties of apples—all in alphabetical order, and all grown organically on Salt Spring. Along the walls were vendors selling apple-related everything, from apple art to apple-scented body lotion.

I quickly found myself at the famous Pie Ladies' booth, where seven of the island's ten pie ladies had baked eighty-three pies for the event, using a range of heritage apples, including Blenheim Orange, Spigold, Alexander, Gravenstein, Cox's Orange Pippin, and Bramley. The women were busy serving the cluster of people,

OPPOSITE:
Apple festival participants get a good look at the more than three hundred varieties of apples on display at Fulford Hall on Salt Spring.
PHOTO: DERRICK LUNDY

who, considering the time of day, must have been making the excellent choice of eating pie for breakfast. One pie lady told me her favourite heritage baking apples are the "lovely" Gravenstein and Bramley. Quickly, however, the other women chimed in and there was a consensus: Wolf River.

"Are they delicious?" I asked, mouth starting to water.

"Not necessarily. They are huge."

And suddenly I got it. They have to peel all these apples—cut out the cores, chop them up. They probably have apple-peeling nightmares. Of course they would like monster apples.

"Sometime you just need one or two Wolf Rivers for a whole pie," they confided.

Another area of the hall drawing a big crowd was the apple identification section, where a long line of people clutching bags of apples formed. Seated at the head of the line-up was Dr. Bob Norton, an eighty-four-year-old apple grower, fruit researcher, and horticulturalist from Seattle, along with some members of the British Columbia Fruit Testers Association, including Clay Whitney, whom I'd met once before, and whom I would see again at the upcoming apple festival at the University of British Columbia.

I had gone to a talk a few nights earlier by Dr. Norton and was amazed by how little I knew about apples and apple-growing. Clay is similarly knowledgeable; he's able to look at an apple and have a pretty good idea of what it might be. I knew I'd be spending a lot more time identifying apples with Clay at UBC, so for today I just hovered, watching him and the others tackle the apples that

landed in front of them. Surrounded by fifteen reference books and a computer, Clay expected fifty to sixty people would line up with their mystery apples at the festival. Inevitably, he added, laughing, someone would arrive with a huge box, just as he's ready to leave.

At the display tables next, I was surprised by the sheer number of apple varieties and the host of different names as well as the many sizes and shapes. There, for example was a red-fleshed crabapple that was the size, shape, and colour of a cherry. Then there was the Mara Red apple, deemed the second sweetest on Salt Spring (Tolman's Sweet got the number one ranking), which was an unusual oblong

First stop for many people at the Salt Spring Apple Festival is the famous Pie Ladies' booth. Seven pie ladies baked eighty-three pies for the event in 2011, using a range of heritage apples, including Blemheim Orange, Spigold, Alexander, Gravenstein, Cox's Orange Pippin, and Bramley.

Apple experts Clay Whitney and Dr. Bob Norton (not shown) staff the apple identification table, viewing, cutting, and tasting the many apples brought in for identification.

shape. Another small corner of the table had apples that varied in colour from a creamy yellow to bright green to a rich, almost magenta red. Salt Spring Fall Fair "winner" notations were attached to varieties of Kings and Gravensteins, as well as a Spokane Beauty, which ironically, was quite misshapen and not really a "beauty" at all. "Best-tasting apple on Salt Spring" was given to the deep-gold-coloured Saint Edmund's Pippin.

Behind the display tables stood clever-looking apple people available to answer questions, even though it was difficult to hear above the din. But I chatted with a fellow who told me he has a Belgian fence (the trees are trained to grow in a lattice pattern) of fourteen apple trees, all organized to ripen at different times. He planted the trees about ten years ago, and prunes and cares for them for about nine months of the year and then harvests the fruits of his labour from August to November.

"When it's in bloom, it looks gorgeous," he said.

I was suitably impressed and when I asked if he ate a lot of apple pie, he confessed that he "likes to eat the pies that other people make. My wife can make wonderful cakes and wonderful cookies but her pies are concrete."

The Fulford Hall portion of the apple festival was really kicking off and it became difficult to move around inside. But things quieted as Captain Apple took the stage, and urged people to get out and go to as many of the sixteen apple orchards on the map as possible.

"The farms are where you really get to connect with the source—the farmer who grows the apple and even the trees that do the growing," he said. "This will connect you with your favourite apple."

A bit of entertainment (apple-related, of course) followed, as an historical island apple grower, James Hector Monk, rose from the grave to talk about apple farming on Salt Spring in the 1800s. Harry farms a part of the original Monk land.

But then it was time for me to get out and visit a few orchards. I planned to grab my eighteen-year-old daughter, Sierra, who would just be crawling out of bed and probably ready for some apple-tasting sustenance. But first, I headed to Ruckle Farm where, among other things, I bought some Wolf Rivers, thoroughly intending to cook my first-ever apple pie. In addition to Ruckle, I had three other "must-do" orchards: Bright Farm, one of the largest apple collections on Salt Spring; Samuel Beddis' historical orchard; and, of course, Harry's farm, which was really the epicentre of the event.

Sierra and I landed at Bright Farm and undertook our first tasting session, a little overwhelmed by the variety of apples available. Bree and Charlie Eagle have 250 apple trees, many of them heritage, set on a 10-acre heritage farm that was once the homestead of an island pioneering family, the Mouats. They had 50 apples set up for tasting and Sierra and I plunged in, immediately discovering that our taste in apples differed. Sierra preferred the softer, mushier varieties, like Wolf River and Chenago Strawberry (New York, 1854), which I thought tasted like a bruise, while I was more partial to the harder, tart-yet-sweet types like Goldgelb, an apple from Germany that had a totally different flavour from the others. We tried Swayzie (New York, 1872); Pomme Gris (Ontario, 1780s), which we agreed was "not bad"; the milder King of Tompkins (also known as King, New Jersey, 1800s); the tart and "pretty good" Spigold (New York, 1962); and the rock-hard Swaar (New York, before 1770), which had an intensely sour aftertaste. We also tried a banana crabapple, which was completely bitter, tart, and unpalatable.

Bree Eagle, who wrote her undergrad thesis on heritage apples, had a map identifying the 250 different trees; she said two of her favourites are Honeygold (US, 1935) and Jonagold (New York, 1953).

After a tour of the orchard itself, we drove to Beddis Castle Orchard, the homestead of Salt Spring pioneer Samuel Beddis. Here we wandered through the truly scenic orchard of 150 trees,

originally created in the 1870s and lovingly restored over the last few decades. At this stop, the item of the day was super sweet and tasty homemade apple juice, which we sampled with enthusiasm. The juice contained all of the orchard's apple types mixed together.

"We put them in big barrels and squeeze until we can't squeeze anymore," said the woman handing out the little cups of juice.

This was actually the previous year's juice stock, frozen and brought out for the early-October festival because the apples wouldn't be ready for pressing for another three weeks. The orchard used to "push" the apples to make the juice in time for the festival, but found the resulting product was tastier when they "waited for the apples to

Captain Apple (Harry Burton) and Salt Spring pioneer James Hector Monk (played by Ken Lee) provide a little entertainment, and give a historical re-enactment during the 2011 Salt Spring Apple Festival. Burton now farms a portion of pioneer Monk's original land.
PHOTO: DERRICK LUNDY

Wasps enjoy a Ginger Gold apple at the tasting table at Apple Luscious Organic Orchard.

PHOTO: SIERRA LUNDY

tell us they are ready." The juice contained nothing but the apples: no sugar necessary! The apples were pressed over a one- or two-day period by juicer Paul Linton, who runs a very well-used apple press throughout the fall on Salt Spring.

Our day was passing and we headed to Apple Luscious orchard, where Captain Apple had laid out over one hundred varieties for visitors to taste. Here we discovered Cox's Orange Pippin. It was described as "one of the finest apples ever grown" and I had to agree. Sierra liked it too.

We fought the wasps on the uber-sweet Ginger Gold, which is a newer apple from a chance seedling on a farm in Virginia, introduced in 1982. We decided the Red Gravenstein was good in its tartness and once again, surprisingly different from the others. We were becoming truly amazed at the variations in taste among the many, many different apple types. Grimes Golden (West Virginia, 1800s) puckered us up with its sour flavour. Also sour was Hidden Rose, one of Harry's collection of red-fleshed apples (also known as Aerlie Red Flesh, from Oregon). Handel's (Turkey, 1800) was bitter; Mother (Massachusetts, 1840), tart; Pink Pearl (Ettersburg, 1940), which makes pink applesauce, was really tasty; and Ribston Pippin was also pretty good.

"Oh, Winekist! I need to try that," I said.

"Of course you do," said Sierra, aware of my fondness for wine.

"Yum, really different. I like that one."

We could have visited many more farms—including cheese-makers and vineyards—all serving up something "apple." I left the festival astonished by the number of apple varieties and tastes as well as the abundance of apple-crazy people on Salt Spring.

But nothing prepared me for the UBC Botanical Garden and Centre for Plant Research's annual apple festival in Vancouver. This time, I picked up my mom for company and we motored over on the ferry, ending up at the university grounds about an hour before the gates actually opened. Nonetheless, hundreds and hundreds of people—young, old, and everything in between—were already lining up, carrying big, empty apple-buying bags.

This was the mid-October festival's 20th anniversary. It takes place over two days every year and is attended by some 14,000 people who purchase over 37,000 pounds of apples and 500 apple trees on dwarfing rootstock. These include more than 70 varieties of heritage apples, the most popular being Grimes Golden, Bramley's Seedling, and Cox's Orange Pippin (of course), and 116 different types of trees—many with names familiar from the Salt Spring festival and many, many new names. The varieties include four crabapples—Chestnut, Dolgo, John Downie, and Kerr—and three cider trees—Brown's Apple, Chisel Jersey, and Kingston Black.

According to a booklet published by the society, the festival got its start in 1991 when members of the Friends of the UBC Botanical Garden (FOGs) sought a way to "draw people to the garden they loved dearly." They decided to celebrate the apple and set out to introduce people to varieties beyond those that can be found in grocery stores.

"By opening day, Oct. 18, 1991, there were thirty-seven hundred pounds [of apples], representing twenty-six varieties, for sale on the deck of the reception centre. Inside, laid out on wicker trays, were forty-one varieties for tasting with ten more just on display. The Friends of the Garden baked and sold apple goodies in the tea corner."[2]

Since that time, the BC Fruit Testers Association has joined the festival, bringing its collection of over two hundred varieties of apples and offering workshops and apple identification. The first event twenty years ago was "modest" compared to today's extravaganza, and as my mother and I wandered through the orchard, I was reminded of attractions like Victoria's famous Butchart Gardens at Christmas time, when it is so bursting with people, it can be difficult to see the garden itself. There were literally people everywhere. On the main lawn, near the apple-tasting tent, at the food fair, selling apple-everything. Line-ups to the booths—apple pie, apple cider, candy apples, dried apples—wound around and into each other. The tasting tent was similarly jam-packed as was the crowd around the children's area and "longest apple peel" contest. (Even I knew that anyone entering the contest should use a Wolf River.)

But what I'd really come for was to check out Clay Whitney and his fellow fruit testers, who were giving demonstrations like cider-pressing, apple grafting, and identification. Surprisingly, the line-up for apple identification here was less daunting than the one on Salt Spring, and I had a chance to watch Clay in action.

Standing in front of him with about seven different bags of apples was a couple who had recently bought an old orchard on Samuel Island, in BC's Gulf Islands. They had nineteen apple trees, seven of which had yet to be identified. Clay had just finished identifying a King apple, which brought the total to seven King trees on their property. Now they handed him an early-ripening apple that Clay immediately thought might be a Yellow Bellflower or Summer Bellflower.

"The suture line," he said, pointing to a narrow, raised line on the outside of the apple, "is really pronounced in some varieties [like the Bellflower]."

He was also tipped off by the long, conical shape of this apple and the fact that it ripened in the middle of August. The taste

Clay Whitney, with the BC Fruit Testers Association, helps a couple from Samuel Island identify apples from their trees at the annual UBC Botanical Garden's apple festival.

was tart and the couple said the apple didn't store very well and needed to be cooked quickly. Clay checked his number one book: *The Apples of New York*, by Spencer Ambrose Beach, published in 1905. He said that although climate and growing conditions can result in minor variations among varieties, this New York-based book, with its photos and descriptions of old apples, continued to be very useful. It also listed apple name synonyms, which is important, because different regions can call apples by different names. Clay also checked a huge file he created on his computer and thumbed through a few of the many other resources he had at his fingertips.

When identifying apples, Clay never likes to say definitively, "This is a . . ." because there are so many possibilities and so many variations among apple types. However, this time, he felt quite certain the apple was one of the two Bellflowers.

"These are becoming harder to find and it's the old orchards you'll find them in," he said. "They fell out of favour because they have no storage qualities."

Usually they taste good, he added, but this one was beyond its prime: the Yellow Bellflower is an early-season dessert apple. It also makes good applesauce, which is what it was ready for now.

To identify the next apple—which he immediately thought was a Roxbury Russet—Clay took me through the entire process. First, looking at the apple's general appearance, including the colour, he immediately determined a number of possibilities (mostly because he'd done this so many times before). This apple had some russeting, a unique characteristic in itself. Of the apple types that have russeting, the pattern differs, so he looked at that too. He also considered the apple's general shape—was it round, oblong, or conical and did it have ridges? Was it flat at both ends? He flipped it over and considered the shape from the bottom. Was it deep? Shallow?

"I'm also looking at how deep it is inside the stem and even the stem itself. This one is short and clubby, another clue that it's a Roxbury Russet. I look at [the stem] from the side and see how far it comes above the base. In a Golden Delicious [for example], it comes right up and out. Here we have a short, clubby stem that goes gently down and deep."

As well, he looked at the lenticils (naturally occurring dots—ventilation pores—on the surface of the skin), which can be "very different from one apple to another." Then he moved on to the inside of the apple, cutting it first horizontally to see the shape of the core. Noting that the seeds are still white, he knew this apple was "not even being close to ripe for another three weeks." He also looked at the shape of the seeds—plump, round, narrow, pointy?—and the shape of the core—were the carpels tight together? In this apple they were separate; in the King he saw earlier, they were wide open. Even fibre bundles in the core and tufting in the seeds gave clues as to the type of apple. Cutting the apple again vertically, Clay looked at the core lines and whether or not the stem was coming straight up from the core or pitching to one side. Once he had all the physical details in mind, combined with information from the apple's owner, such as when it ripens and how well it stores, he tasted it, considering flavour and texture. This apple was a little tart, we discovered.

Then it was time to hit the reference books, of which he owns about two hundred; he had just under twenty with him. He was a bit stumped because one reference said the Roxbury Russet has a long stem, and we were looking at a short, clubby stem. So he ruled out Roxbury Russet for the moment.

"And I'm also going to rule out a Golden Russet because it's sweet [and this one is tart]. And it doesn't have the shape of a Hudson's Golden."

Then he went back to Beach's book and found that contrary to the other reference, this description of the Roxbury Russet said the stem is "short to medium, thick or swollen."

All the other characteristics matched and it appeared Clay's initial intuition was correct. "I always say 'possibly,'" he stressed again, but he believed this was a Roxbury Russet. Clay went on to identify a Winter Banana from the couple's orchard, although he was puzzled at first by the apple's delicious taste. "I don't like them," he said. "I've never had a good one."

I left this festival, like the Salt Spring festival, with apples on the brain—as well as a new appreciation for the tastes and varieties of apples of the past.

PHOTO: CLAY WHITNEY

Taste and appearance: A greyish-green russet apple. The yellow-green flesh is firm and coarse-textured, with a crisp and tart taste.

Use: Eating fresh, cooking, and making juice and cider.

History: It was first discovered and named in the former town of Roxbury, part of the Massachusetts Bay Colony, which is now part of Boston.

Growing and harvesting: The tree is described as vigorous, scab resistant, and productive. Picked in mid-October, the apple will keep until spring.

Other: Once the most popular russet apple, it lost its popularity because of its unattractive appearance, and the general commercial dislike of russet apples.

WINTER BANANA
INDIANA, 1876

PHOTO: CLAY WHITNEY

Taste and appearance: Beautiful in appearance, with a pale yellow skin and faint pink blush. Described by one source as mild-flavoured, and by another as "sweet and tart, with a definite banana aroma."

Use: Eating fresh and cooking.

History: It originated in Indiana in about 1876.

Growing and harvesting: Picked in mid- to late October, it keeps until December, but can become "mealy" if stored too long.

Other: One source suggests its dense texture makes it better to eat when sliced and paired with cheese.

WOLF RIVER
WISCONSIN, 1875

Taste and appearance: Noted for its large size, the apple is flushed dark red, with white, soft, slightly tart, juicy flesh.

Use: Cooking—it keeps its shape when cooked and, being fairly sweet, doesn't need much sugar. Often only takes one apple for an entire pie.

History: It was discovered near Wolf River, Wisconsin, in about 1875 from an open-pollinated Alexander.

Growing and harvesting: The tree is very cold hardy (making it a good choice for growing in the northern part of North America) and has a natural resistance to scab, fireblight, and mildew. It is picked in mid-September and lasts in cold storage until December.

Other: Wolf River can be very large; apples weighing a pound are common.

RIBSTON PIPPIN
FRANCE, 1688

PHOTO: CLAY WHITNEY

Taste and appearance: Slightly russeted with dark red streaks over a red-gold flush. It has firm, dry but crisp, yellow flesh with an intense, robust, and sweet flavour.

Use: Eating fresh, cooking, and juice.

History: It was raised from seed brought from Rouen, France, about 1688 and planted in Ribston Hall in Yorkshire, where it was first grown in in the eighteenth century.

Growing and harvesting: Picked in early October, the apple improves in flavour with storage but passes its prime by January. The flavour is at its best a month or so after picking.

Other: Much of the interest around Ribston Pippin is due to its parentage of the more famous Cox's Orange Pippin.

Apple Crumble

1 cup oatmeal

1 cup flour

1 cup brown sugar

½ cup margarine or butter

2–4 apples

raisins or blackberries (optional)

Combine oatmeal, flour, and brown sugar. Blend in margarine or butter until crumbly. Peel and slice apples and place in the bottom of a loaf pan. Alexander and Wolf River apples will make a drier crisp; Gravenstein will make a juicier crisp. If using sour apples adjust how much sugar you add with dash of lemon juice. Raisins are optional. Or, add a few blackberries for variation. Spread crumble mixture on top of fruit. Bake at 350 degrees for about 30 minutes until apples are soft and crumble is slightly browned.

Recipe provided by Marjorie Lane of Ruckle Farm

Apple Pie

Filling:

8 medium-sized apples (use Granny Smith and
   mix with other varieties)

⅓–⅔ cup sugar

¼ cup all-purpose flour

½ tsp ground nutmeg

½ tsp ground cinnamon

pinch of salt

2 Tbsp butter (for use before baking)

Crust (recipe makes 1 double crust):

2½ cups white flour

2 Tbsp sugar

¼ tsp salt

½ cup cold butter, broken into small pieces

5 Tbsp cold vegetable shortening

8 Tbsp ice water

For the filling—peel, core, and slice apples. Try to keep the size of the slices even. Mix sugar, flour, nutmeg, cinnamon, and salt in a large bowl. Stir in apples.

For the crust—measure flour, sugar, and salt in a large bowl, and stir to combine. Add the chilled butter pieces and shortening to the bowl. Cut them in with a pastry cutter or knife. Don't overmix. Add ice water and mix until the dough holds together (add more water if necessary). Turn the dough onto a lightly floured surface, knead it together, and then divide it in half. Flatten each half into a disc, wrap in plastic wrap, and chill for at least half an hour. Roll out one of the discs on a lightly floured surface until you have a circle that's about 12 inches in diameter. Put the circle in a 9-inch pie plate, trimming any extra dough from the edges with a sharp knife. Return it to the refrigerator until you are ready to make the pie.

Heat the oven to 425 degrees. Pour filling into pastry-lined pie plate. Dot with butter. Roll out the second ball of dough and cover top. Use a fork or your fingers to pinch the edges together. Cut a couple of slits in the top. Optional: cover edge with a 3-inch strip of aluminum foil to prevent too much browning. Remove foil during the last 15 minutes of baking. Bake for 40 to 50 minutes or until crust is brown and juice begins to bubble through slits in crust.

Recipe provided by Steve Glavicich, chef/owner, Braizen Food Truck, Calgary

CHAPTER FIVE

Biting Into a Surprise

One of Harry Burton's
red-fleshed apples
called Grenadine.
PHOTO: RON WATTS

Apple Facts

To get a good apple crop, 6 to 15 per cent of the blossoms must be pollinated.

Apples float because 25 per cent of their weight is air. Pears sink.

Sliced apples turn brown with exposure to air due to the conversion of natural phenolic substances into melanin upon exposure to oxygen. Sliced fruit can be treated with acidulated water (water with some sort of acid added, such as lemon juice, lime juice, or vinegar) to prevent this.

Apples ripen six to ten times faster at room temperature than if they are refrigerated.

On a sunny-but-crisp day in April, I arrived at Harry Burton's Apple Luscious Organic Orchard, just down the road from my home on Salt Spring. Trees were bursting with blooms, plump chickens scampered about, and I could practically feel the vibration of bees humming their paths from one plush blossom to the next. The earth was muddy and Harry, shovel in hand, was clad in knee-high gumboots and a set of work clothes comfortably past their "best-before" date. As if determined to match the busyness of the bees, he was a picture of activity, shovelling, lifting, and happily talking about apples. Harry knows a lot about apples and he's called on regularly to answer questions. Frequently asked, "What's your favourite apple?" he always answers, "The one I'm eating at this moment—they are all good." But given the choice of apple topics, he'll extol the glories of red-fleshed apples, which he believes will be the next wave of apple excitement. Growing and experimenting with twenty-four varieties, Harry almost certainly has the largest collection of red-fleshed apples in Canada.

On that day, I tromped behind him alongside a fellow called Charlie Van Streubenzee, who was picking up five heritage apple seedlings that he had ordered from Harry the previous year. The young trees—really just three-foot-tall stems with four outward-pushing branches—were a year old and would not produce apples for another four years. Charlie was buying heritage apple trees because he found them "more interesting" than commonplace varieties. He was less concerned about the type of apple tree than the fact he was getting it directly from the grower, especially someone as "keen" as Harry. This undoubtedly made Harry happy because connecting farmer with buyer is one of his goals. He also aims to "get people excited about apples again" and encourages residents to buy locally grown and organic food.

Charlie's purchases were the last of Harry's tree sales for the season. Next year's trees, about eight inches high and all tagged according to variety, sat lined up in tight rows like a block of soldiers. There were six hundred seedlings there, all recently grafted and planted.

Charlie had two young children and before he left, new trees in hand, Harry said, "Next year try a red-fleshed apple tree. Every kid has to have a red-fleshed apple."

After that first visit in April, I dropped by to talk to Harry or sample apples three or four times. Each visit, I followed him around as he tirelessly dug holes, pushed wheelbarrows, shooed chickens from underfoot, and threw oyster shells or straw at the base of trees. Obviously, caring for an orchard of this size keeps Harry busy: there's grafting, planting, transplanting, pruning, nurturing, harvesting, and selling what amounts to eight thousand pounds of apples and some two hundred seedlings each year.

Harry has three hundred trees, translating into over two hundred varieties of certified organic heritage and new apples. Taste is the most important characteristic—"If it tastes good we have it." Harry experiments with different varieties, referring to his in part as a "test orchard," and noting that trees that grow well in the interior of BC don't necessarily do well on the West Coast. He culls trees with less flavourful apples.

Red-fleshed Pink
Delight apples lined
up on a railing.
PHOTO: KAREN MOUAT

"I loved the idea of growing good, healthy food," said Harry, who, during his years as a professor of environmental protection at Canadore College in Ontario, specialized in hydrology, air and water monitoring, and environmental management.

Harry's land comes with its own bit of apple history as it's part of one hundred and sixty acres that pioneer James Hector Monk bought in 1904. Monk grew apples on the land for forty years, alongside his contemporary Henry Ruckle. Although none of Monk's trees exist on Harry's corner of the property, he grows some of the varieties—like Gravensteins, Baldwins, and Kings—that once filled Monk's boxes of apples, which sold for two dollars apiece.

Harry doesn't necessarily favour heritage apples over newer apples (again, it's the taste thing), but he does see a revival of interest in older varieties, especially as part of the Slow Food Movement, and a coinciding interest in the origin of the food people eat. Slow Food—as opposed to fast food—is an international movement that aims to preserve the growing, purchasing, and eating of local food. Heritage apples fit nicely into this cause.

"What we have now is the cream of the cream," said Harry. "The heritage apples that are still around are very tough and good apples. In other words, someone kept them going over the years. If they were weak, they wouldn't have survived."

Prized among Harry's heritage and new varieties are the red-fleshed apples, including some descendants of the first red-fleshed apple to land in North America called—rather appropriately—Surprise. Harry's collection doesn't include Surprise, because what he once thought was Surprise turned out to be a different red-fleshed apple called Sirprise. But it is these varieties that truly pluck his heartstrings, and he wants to "to grow every good-tasting red-fleshed apple variety possible."

"People go crazy for them. That's my biggest order—for red-fleshed trees," said Harry, as once again, he busily worked away in the orchard, this time replanting a red-fleshed Pink Princess tree into a spot where two red-fleshed Pink Pearls hadn't thrived. (Pink Princess, Harry explained, was originally called Pink Lady by the late Fred Jansen, a highly esteemed heritage grower who discovered it in Ontario in the 1960s. However, a white-fleshed apple in

OPPOSITE:
A vibrant-coloured
Hansen red-fleshed
apple sits on the left,
next to a Hidden
Rose on the right.
PHOTO: RON WATTS

Australia had already been patented under that name, so the "lady" was upgraded to "princess.")

He believes interest in red-fleshed apples is about to explode. "They are incredible to eat, great for your health, irresistible to children, and beautiful to look at. They are tastier, prettier, unique, and uncommon."

The anthocyanins responsible for the apples' red colour—which ranges from a deep, almost beet-red to a light pink flush—give the apples a nutritional boost and add an interesting twist to apple dishes since the pigmentation doesn't disappear during cooking. (The red colour is lost in fermentation, however, so red-fleshed apples don't create pink cider.) Chefs are among Harry's best buyers, using the red-fleshed apples for everything from pink strudel to pink apple pie, and his "reds" are the first of his apples to sell out each year.

"I don't mean they will take over the apple industry, but they will be such a neat addition to our current brands. Think of pink grapefruit and how popular it's become."

Red-fleshed apples do not necessarily have red skin—like the Hidden Rose I sampled, which was a gnarly-looking green apple with a tart-but-tasty flavour. In fact, there are two different styles of red-fleshed apples. One is red from the time the apple forms: the wood is red, the leaves are red, and the bark has a dark sheen to it. These include varieties such as Scarlet Surprise and Winekist. The second type of red-fleshed apples, like Pink Pearmain and Hidden Rose, don't "red up" until the last three weeks.

"If you pick them early, they're not going to be red. They have no sign at all that they are red-fleshed—the wood is white and the leaves are green. The only one thing the two types have in common is a pink blossom. All other apple blossoms are white."

Among Harry's biggest red-fleshed sellers are the ones that ripen first, at the end of August, like Scarlet Surprise, Winekist, and Geneva 163. But they all sell well in their own time period. Despite this interest, "no one else is really doing them in Canada. Over the years, there has been a lot of resistance because growers are generally traditionalists."

There have also been problems with taste—although Harry's twenty-four varieties taste good or they wouldn't be there—and some difficulty with growing. According to the apple resource website Orange Pippin, "many of them grow poorly, have below-average disease resistance [and] seem prone to biennial bearing—fruiting every other year rather than annually." [1]

Harry agrees it may be "partly true that some are weaker." In the decade he's been experimenting with red-fleshed apples, he's replanted and started over with some varieties. But he believes over time, they will become "good and strong," and he disagrees wholeheartedly that taste is an issue. "People who talk about the taste of red-fleshed apples don't really know, because they don't have the good ones. I know how incredible they are."

Albert Etter (1872–1950) is considered the "father of red-fleshed apples" and many of the red varieties found today were developed in his apple-breeding program at Ettersburg, California, in the early to mid-1900s. One of the most informative accounts about Etter, as well as his fellow red-fleshed apple propagator of the same era, Dr. Nels Hansen (1866–1950), appears in an article, "Albert Etter and the Pink Fleshed Daughters of Surprise," by Ram Fishman.[2] Ram and his wife, Marissa, run Greenmantle Nurseries, near Ettersburg in Garberville, California, from which they have reintroduced Etter's Thornberry and Pink Pearmain varieties, among others.

The two genealogical strands of red-fleshed apples available today are both most likely traced to a crabapple called Niedzwetzkyana found growing in Turkestan. Upon its discovery, this apple drew immediate attention for the extreme purplish-red colour found in its skin, flesh, seeds, foliage, blossoms, and bark. Today, varieties descended from Niedzwetzkyana tend to have darker, deeper colours than other red-fleshed varieties. Nels Hansen worked with red-fleshed apples descended directly from Niedzwetzkyana.

Etter, on the other hand, developed his red-fleshed varieties from Surprise, an apple that also probably descended from Niedzwetzkyana, but appeared in North America via German immigrants in the mid-1840s. Surprise had found its way from Russia to England and Germany by the early 1800s, but the apple-eating public did not embrace it. It was pale green and small; the "surprise" occurred when people bit into it and discovered its pink flesh. The taste, however, was said to be acidic and tart. In North America, Surprise was "scorned" by some and "prized" by others. Pomologist Charles Downing (1802–1885) had a Surprise tree in his collection, but gave the apple poor reviews. However, as a teenager, soon-to-be-famous horticulturalist Liberty Hyde Bailey (1858–1954) became passionate about it, and grafted scionwood from one of Downing's Surprise onto a tree in his father's orchard in Michigan. A single Surprise tree was found still standing on that land—which had been converted into a hospital site—in 1957.

Later, another young man became smitten with Surprise. Etter, a self-taught fruit breeder (who was known even more for his work with strawberries), obtained six hundred varieties of apples from the University of California and set up a breeding program at Ettersburg. Etter used Surprise as the genetic foundation for new varieties of red-fleshed dessert apples. He eventually proved that the longer growing season in the west resulted in a tastier, more attractive version of Surprise than the one dismissed by Charles Downing. By 1944, Etter claimed to have thirty new varieties; however, only one, Pink Pearl, was selected for patenting. The others were mostly ignored and left to deteriorate in his orchard until Greenmantle Nurseries began a rescue mission in the 1970s.

Hansen began experimenting with red-fleshed apples around the same time as Etter. He was on a botanizing expedition to Russia in 1897 when he came across the recently discovered Niedzwetzkyana. He managed to import Niedzwetzkyana scionwood and established a tree in South Dakota, where he lived. A crabapple, Niedzwetzkyana is barely edible—astringent and sour. However, Hansen discovered that when it was crossed with more common apples, the offspring tasted better and still retained the red colour, albeit a little lighter. He eventually introduced several red-fleshed apples, including the

red-wine-coloured Almata, which Harry carries at Apple Luscious. (As Harry points out, Etter is generally considered the winner over Hansen in the "red-fleshed race," but his California climate was much less harsh than Hansen's in South Dakota.) In varieties directly descended from Niedzwetzkyana, the entire tree usually has the red pigmentation (leaves, bark, roots). But the taste is poor and they're often used for juicing or cooking. Almata is an exception, agree Harry and Orange Pippin, in that the flavour is "reasonably good. There is a good quantity of juice, some sweetness balancing a light acidity, and the flesh is soft with some crunch to it."

(Research on Niedzwetzkyana brought up an interesting aside. According to the Global Trees Campaign, Niedzwetzkyana was

Red-fleshed Winekist apples, from Apple Luscious Organic Orchard, are red from the inside out.
PHOTO: HARRY BURTON

identified in 2006 as one of the world's most threatened apple species by the International Union for Conservation of Nature: "It occurs at very low densities and, given the extent of fruit and nut forest destruction, populations are thought to be severely reduced and highly fragmented. Surveys during 2007 found just thirty-nine adult trees in the rapidly shrinking fruit and nut forests of Kyrgyzstan; its population in neighbouring countries is unknown.")[3]

Red-fleshed apples have become the subject of studies in several countries (adding strength to Harry's assertion that they are the wave of the future), including the government-owned Plant and Food Research company in Auckland, New Zealand. Here, plant molecular biologist Richard Epsley is among those working to create a red-fleshed variety that will be popular among consumers. In a YouTube video,[4] Epsley discusses the benefits of red-fleshed apples and the work being undertaken to breed a consumer-popular version. The video is interesting, especially from a scientific standpoint, but it can also be seen as a microcosm of what has gone right—and wrong—with apple breeding through the ages.

Epsley says red-fleshed apples are more nutritious than their white-fleshed counterparts because the anthocyanins are antioxidants that can protect against ailments such as heart disease and cancer. As Epsley sits, holding a deep-red-hued apple from Turkestan, he says consumers can't eat the red-fleshed apples because they "taste awful." (I could practically hear Harry's sigh.) His program works with apple breeders to combine these healthy red anthocyanins with apples that have consumer-accepted taste and appearance. But finding the perfect new apple cultivar via apple breeding can be a decades-long endeavour, so the research centre set about identifying the gene responsible for the red pigmentation. Eventually, it discovered the gene (MYB10), and now, in the race to create a new "delicious" red-fleshed apple, breeders are able to track the gene in progeny.

"They're trying to reinvent the wheel," Harry points out in response, since so many good-tasting varieties already exist. "But don't get me wrong—I welcome the research. The more red-fleshed apples they bring out, the better."

So it seems very likely that there will soon be consumer-accessible (in taste and appearance) red-fleshed apples available on the market. But as these futuristic brands of red-fleshed apples become popular, the threat to heritage varieties increases—making the work of heritage apple growers like Harry, who continue to preserve gene diversity, even more important. And in the meantime, weirdly juxtaposed against all the gene-searching and carefully monitored apple-breeding programs, stands Harry in his orchard at Apple Luscious on little Salt Spring Island, gently replanting a Pink Princess and talking about his growing collection of good-tasting red-fleshed apples, which sell out every season.

ALMATA
SOUTH DAKOTA,
1942

PHOTO: RON WATTS

Taste and appearance: Red skin and red flesh. A good quantity of juice, tart with some sweetness, and the flesh is soft with some crunch to it.

Use: Eating fresh and cooking. (Described as good for colourful applesauce and jelly, and useful in pickling.)

History: Developed by Dr. Nels Hansen in South Dakota, mid-1900s, a descendent of a crabapple called Niedzwetzkyana.

Growing and harvesting: Ripens in late July to August.

WINEKIST
USA, MID-1900s

Taste and appearance: Dark red leaves, blossoms are red. Wood is pink when you cut into it. The fruit has wine-red skin with areas of darker red and very small white dots. Very juicy coarse flesh is almost solid beet-red.

Use: A good addition to cider or sauce.

History: Developed by orchardist and variety collector Morris Towle (1911–1993).

Growing and harvesting: Season is early to mid-August.

PINK PEARL
ETTERSBURG,
CALIFORNIA, 1940

PHOTO COURTESY APPLE LUSCIOUS ORGANIC ORCHARD

Taste and appearance: Medium-sized with cream and light green skin blushed with a red cheek. Bright pink flesh with a rich, sweet-tart flavour.

Use: Eating fresh and cooking, especially applesauce and pies.

History: Developed by Albert Etter, a seedling of Surprise.

Growing and harvesting: Ripens in late August to September.

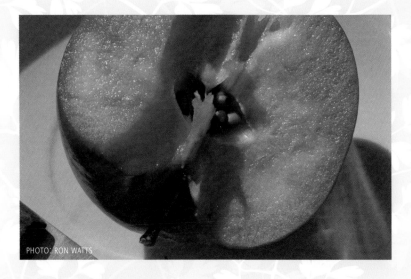

PHOTO: RON WATTS

PINK PEARMAIN
ETTERSBURG,
CALIFORNIA, 1940s

Taste and appearance: Deep-red-striped skin covering vivid mottled pink flesh. Tart with an aromatic flavour.

Use: Eating fresh and cooking.

History: Found growing in an old orchard near Whale Gulch, California, by Ram and Marissa Fishman of Greenmantle Nursery, and thought to be an Albert Etter-developed apple.

Growing and harvesting: Ripens in early September and must be fully ripe for maximum sweetness and colour.

Other: Intense heat in August and September can adversely affect texture and flesh colour. Pink Pearmain is a registered trademark of Greenmantle Nursery.

French Apple Clafouti

4 cups peeled, sliced, red-fleshed apples

1½ cups whole milk

4 eggs

½ cup all-purpose flour, sifted

¼ cup sugar

1 tsp vanilla

1 tsp Calvados or apple brandy

Preheat oven to 350 degrees. Lightly grease a deep 10-inch pie plate. Arrange apples evenly over the bottom of the dish. Combine milk and eggs in a blender until smooth. Add remaining ingredients and blend for 5 seconds. Pour batter over apples. Bake for 1 hour, or until a toothpick inserted in the middle comes out dry. Serve warm.

Recipe provided by Steve Glavicich, chef/owner, Braizen Food Truck, Calgary

Red Apple Preserves

8 large red-fleshed apples

1 cup water

1 Tbsp lemon juice

1 package powdered pectin

4 cups sugar

2 tsp ground cinnamon

large pot and boiling water canner rack

Peel and core apples. Slice each apple into wedges. Combine apples, water, and lemon juice in the large pot. Cover the pot and simmer for 10 minutes. Stir in the pectin and bring the mixture to a rolling boil. Continue stirring frequently. Add the sugar and boil for another minute, continuing to stir frequently. Remove the pot from heat and add the cinnamon.

Wash the empty jars and lids in hot, soapy water. Place the jars and lids on the boiling water canner rack. Submerge the rack in boiling water for 10 minutes. Empty each jar of water and immediately pour in the hot preserves. Do not allow the jars to cool before using them. Fill the jars within one quarter of an inch of the lid. Wipe any spillage from the sides of the jar and screw on the lid. Place the filled jars back on the boiling water canner rack. Submerge the preserves in boiling water for 5 to 15 minutes. Remove the jars and place on a towel to cool.

Recipe provided by Steve Glavicich, chef/owner, Braizen Food Truck, Calgary

CHAPTER SIX
"Surely the Noblest of All Fruit"

Sooke, on Vancouver Island, is home to dozens of heritage apple trees, some planted by Captain Walter Colquhoun Grant, a Scotsman who landed in Sooke in 1849. (Grant is also credited with introducing the now infamous Scotch broom to North America.)

Apple expert and enthusiast Clay Whitney has dedicated much of his life to the preservation of heritage apples trees, like this one at a historic site in Sooke.

Apple Facts

Facts gathered by the University of Illinois:[1]

Apples harvested from an average tree can fill twenty boxes that weigh forty-two pounds each.

The largest apple picked weighed three pounds.

America's longest-lived apple tree was reportedly planted in 1647 by Peter Stuyvesant in his Manhattan orchard and was still bearing fruit when a derailed train struck it in 1866.

One of George Washington's hobbies was pruning his apple trees.

The irony became obvious as I met up again with apple expert Clay Whitney, this time in the Sunriver Community & Allotment Gardens in Sooke, on Vancouver Island.

It was a beautiful, sunny fall day and the community garden, with its nicely organized plots of soil, brimmed with leafy greenery. Clay showed me a string of young fruit trees and another line of apple trees, recently grafted from some of Sooke's heritage trees. On a ridge above the community garden sat a ring of new houses, recently built as part of a huge development called Sunriver Estates. Across the street, a pretty eleven-acre parcel was all that remained of a heritage apple orchard planted in the mid-1800s.

From where we stood in the garden, we could see in front of us the newly planted row of narrow, stick-like young apple trees—still years away from producing fruit—while across the road towered two massive, hundred-year-old King trees at the edge of the old orchard. Clay said the back end of the property contained several Baldwins and a Lemon Pippin or two, and he estimated these trees produced about six hundred pounds of apples per year, many of which fed the bears that had visited the land for decades. Clay feared that whoever bought this property would take down the heritage trees to make way for development, and also to keep out the bears.

How ironic that fruit trees had been planted in a garden—created in part to provide food for the community—while across the street, old trees producing barrels of fruit could face a chainsaw.

Preserving the old apples trees in Sooke has become a passion for Clay, a landscaper, who commutes among Sooke, Victoria, Prince George (where his wife, Sierra, attends medical school), and various jobs all over southern Vancouver Island. In addition to busily pruning fruit trees, grapes, and berries, Clay lectures to garden clubs, travels the apple festival circuit, and has had inquiries from a group of people on the prairies, wanting help with planning and planting orchards there. Much of his apple work—like that in Sooke—is volunteer.

Clay believes the remaining Lemon Pippins planted in Sooke in the mid-1800s are some of the oldest apple trees in the province. His "call to arms" occurred in January 2010, as he headed to a farm on part of the original homestead of Captain Walter Colquhoun Grant, who landed in Sooke from Scotland in 1849 and carved an estate called Mullachard out of the wilderness. Aiming to get scionwood from the three huge, 160-year-old apple trees left in the orchard, Clay quickly discovered they'd been taken down by a new owner to make way for a subdivision.

"I came around the corner and saw a big open space—I started feeling sick—it looked like the moon," he recalled, adding that the trees were healthy and producing about 150 pounds of apples a year. "People can't always put that into context, but compare it to a 5-pound bag at the grocery store and that's a lot of apples."

Clay asked a nearby backhoe driver if the uprooted trees were still on the property because "even ripped out, the scionwood would have still been usable." However, they were gone.

"Those trees were still alive and producing food, and nobody took the time to graft up more trees. If I'd known they were coming down, I would have done it sooner and got the scionwood."

The loss of those trees prompted Clay to write to his local mayor, MLA, and MP, drawing their attention to the disappearance of trees and the corresponding link to Sooke's history and culture. He also stressed his belief in the importance of maintaining and producing locally grown food.

"I pointed out how important it is to save what we have. It's difficult to import anything, so what we've got is what we've got. We have to keep it for its historical and genetic importance."

Canadian federal government regulations limit the importing of apple trees and scions so it is difficult to reintroduce some varieties once they are lost.

"If the material is coming from a source that has been federally inspected and has the proper paperwork, then it is allowed into Canada," Clay said. "But most of the small growers who grow rare and obscure varieties don't have these credentials, and it is too much work for them to do it for the number of scions or trees they sell to Canadians."

Stung by the loss of those trees, Clay forged an idea to graft as many of Sooke's heritage trees as possible and plant them in the Sunriver Community & Allotment Gardens. And in response to his correspondence, the District of Sooke developed a plan to create a database that would flag the locations of heritage trees should a property sell or an owner apply to undertake a major home renovation. Unlike some municipalities, Clay said, Sooke does not have a

These red-fleshed Pink Delight apples show a natural, waxy shine.
PHOTO: HARRY BURTON

heritage tree bylaw and, although he is pleased with the steps being taken, he remains concerned that some trees could fall through the bureaucratic cracks.

Before our stop at the Sunriver Community & Allotment Gardens, we visited Woodside Farm and its 150-year-old fruit trees, including a towering Lemon Pippin, a Gravenstein, and a Esopus Spitzenburg (famous as American president Thomas Jefferson's favourite apple). Woodside Farm contains one of the two remaining houses built in 1884 by John and Ann Muir, who obtained Captain Grant's holdings when he returned to Britain in 1853. (Although Grant is credited with planting some of the oldest apple trees in Sooke, he is even more renowned for introducing Scotch broom, the invasive plant that now grows from Alaska to California, to the area. He said the bright yellow flowers reminded him of home.)

The apple trees at Woodside Farm are set against a peaceful backdrop of grassland and farm animals, including a gaggle of squawking fowl, and, of course, the charming old farmhouse itself. Clay showed me an old Lemon Pippin that had been split in half—likely due to a weak crotch in the tree—and fallen, but had completely re-rooted itself at both ends, each now with a mass of trunks and branches growing above it, all laden with fruit.

"Just because it's fallen over doesn't mean it can't still produce apples," Clay pointed out, adding that when he holds pruning workshops at the farm, he always directs people's attention to this tree.

Down the road, we stopped at a property where a huge apple tree soared up and beyond the hydro pole. Clay wanted to measure its circumference because he believed it to be one of the biggest apple trees he's ever seen. He estimated it's 140 years old, 40 feet tall, and, as it turns out, 7 feet in diameter. Although the owner called it a Yellow Pippin, he was certain it's a Lemon Pippin.

Our final stop was the Sunriver Community & Allotment Gardens—a project of which Sooke has every right to be proud. Located just a few kilometres off the main road, the 2.5-acre garden was created in 2010 through the Sooke Food Community Health Initiative Society with the goal of building awareness about local agriculture and food sustainability. Helpfully surrounded by deer fencing and rabbit wire, the garden's sixty plots are available to rent for fifty dollars per season. In addition to the individual plots, there are community plots and spaces to hold workshops on pruning styles and techniques.

Following its creation, the garden underwent a second phase of development with the planting of the orchard the following spring. Clay, with two other local volunteers—horticulturalist Glen Thelin and Bonnie Jones—surveyed, sketched, and landscaped the northern part of the gardens and then planted fifty-eight fruit trees, including apple, pear, plum, and cherry. Also in the works was Clay's "baby," the creation of the Heritage Fruit Tree and Demonstration Orchard with grafts from local heritage trees. When I visited in the fall of 2011, Clay had already donated twenty grafts of the trees planted by pioneers, as well as a few of his own favourites. He had also donated fourteen trees to the Ladybug Garden, run by the region's T'Sou-ke Nation, who are considered environmental leaders among Canada's aboriginal peoples. Among their projects is a nationally acclaimed solar power program through which they generate energy and sell it back to the grid. The band is also working toward food sustainability by producing much of its own food in its garden and greenhouses.

"I donated a bunch of trees to them because I admire them so much," Clay said. "These trees are just varieties I enjoy, but as we were moving so far [to Prince George], I couldn't take them with me. I really felt that the T'Sou-ke would appreciate them and take care of them—and they have. I plan on giving more to the T'Sou-ke once I graft some more of the local trees."

Clay passionately believes we need to "take back the land for our food." He recently told a woman buying trees from him that, ten years from now, this could be the most important money she's spent. "We need to learn to take care of our lands and grow our own food—but we have forgotten how. We've lost the basics."

Sooke, it seems, is one of the communities in Canada working to rectify this. In fact, judges in a national town-enhancement program called Communities in Bloom said in their 2011 report that Sooke could easily become the eco-capital of Canada. The report also praised the T'Sou-ke Nation and the allotment gardens.

"The community garden is an amazing place . . . In keeping with the keen interest of citizens in producing local food, the establishment of a heritage fruit tree program is exciting. The orchard in the community garden will become a place of interest to those wanting to learn about heritage fruit trees." [2]

Other groups, societies, and communities across Canada are working to preserve and educate people about heritage apples and other fruits. Much of Clay's volunteer work—like the hours he spends at various festivals helping to identify apples—is through the British Columbia Fruit Testers Association, "an independent and all-volunteer group of fruit-growing enthusiasts." The group includes hobby growers, commercial orchardists, professionals, and others who have a "common interest in the science and practice of fruit cultivation." Among the association's goals is to "identify and preserve heritage fruit varieties that are part of our agricultural heritage. By encouraging their propagation, we preserve their genetic diversity for future generations."[3]

The group holds regular grafting and pruning workshops, and publishes a newsletter, *The Cider Press*, four times a year.

Seeds of Diversity is another Canadian organization working to preserve heritage plants. A project launched by the Canadian Organic Growers in 1984, Seeds of Diversity is now an independent charitable organization, "dedicated to the conservation, documentation and use of public-domain, non-hybrid plants of Canadian significance." It has fourteen hundred members from across the country and "grows, propagates and distributes over twenty-nine hundred varieties of vegetables, fruit, grains, flowers, and herbs."

"We are a living gene bank," the website notes. On its online heritage plants database, it includes descriptions, stories, history, and cultivation details, plus gardeners' comments on nineteen

Clay Whitney measures what turns out to be one of the largest apple trees he's ever seen. Located on a property in Sooke, it's 7 feet in diameter, 40 feet tall, and over 140 years old.

thousand cultivars of Canadian garden vegetables, fruit, grains, and ornamentals, including ninety-four apple varieties.[4]

Another hotbed of heritage apples is located in Creemore, Ontario, in the Georgian Bay area. Here, the Creemore Heritage Apple Society is the umbrella for more than 10 heritage apple orchards, representing 250 varieties of apples. The society has two annual events: a meeting in the spring and (apparently even more popular) a highly competitive apple pie contest held every Thanksgiving. The website asks, "Why heritage apples?" and answers, "Because extinction is forever and the move to mono-culture farming endangers the survival of many of the old apple varieties . . . Perhaps the best reason for the society is that many of the old heritage apples simply surprise you [with their] great taste."[5]

One Creemore grower speaks of the link heritage apples have to past generations, "Ask your parents or grandparents . . . what their favorite apple was, and my goal was to have their apple in my orchard."[6]

It is this link that resonates with so many people, including Clay, who has for the last several years been attempting to graft an apple tree from The House of All Sorts, where the painter Emily Carr lived in Victoria. Carr built the "all sorts" boarding house in 1913 adjacent to Emily Carr House, where she spent most of her childhood. Clay believes Emily Carr planted this tree.

"I had posted an ad on Used Victoria for anyone who had too much fruit, saying that I'd harvest it in return for some apples—this was before LifeCycles really got going," Clay recalled. "Over the phone, the owner of the house mentioned that the tree was a seedling planted by Emily Carr. When I got there I found this thirty-foot tree with a circumference of around five feet. This would definitely be the size if it was planted in Emily Carr's time." The tree looks like an Alexander but Clay said the apple is a slightly different in taste and "has small variable attributes like stem, calyx, and cavity. I can't remember where I saw it, but I remember reading that Emily's favourite apple was an Alexander." If Emily Carr did plant it as a seedling, "it's the only tree in the world of that variety." He named it Emily's Pippin and has been "desperately" and unsuccessfully trying to get the tree to propagate for several years.

Genetic biodiversity is another of the reasons Clay believes we need to save heritage apple trees. These trees are probably better able to resist disease such as canker, which he says is rampant. Apple canker is a fungal disease that attacks the bark, causing a sunken area of dead bark and eventually the death of the branch. Sometimes the canker stabilizes or heals but more often, it spreads gradually until the branch is girdled, or so weakened that it breaks.

"Canker is everywhere. It is in our atmosphere and the spores are part of what forms the nuclei of our precipitation. There has been so much importation over the years that before many regulations were in place, the various forms of cankers were being spread all over the earth," he said, also attributing its proliferation to poor pruning practices. "With the lack of pruning knowledge I see everywhere I go—the way people hack their trees in fall and winter—we are just creating the perfect environment for this disease to spread

Seedlings growing at the Sunriver Community & Allotment Gardens in Sooke, BC, where a community garden has been created amid a new subdivision.

unchecked. It makes me want to cry every time I hear someone say they heard or read that you only prune fruit trees in winter."

Clay firmly believes that summertime pruning is essential to limit the spread of canker and explained his reasons in an email: "Auxins are concentrated at the tip of each branch . . . and act as a brake for the buds below, preventing them from shooting up. When you make heading cuts in the winter, you remove the branch tips where the auxins are concentrated—and also remove the brake causing vigorous growth when the tree comes out of dormancy . . . Each one of those wounds is now a possible infection point from canker spores, which are being produced in late January through May. So I do very little or no dormant pruning until mid-March when the tree is actually waking up, able to start healing its wounds and not be exposed for such a long time. When you make heading cuts in summer (roughly July 15 to August 7) the tree will put on very little growth before going into dormancy. During this time, auxins again become concentrated at the growing tip. As long as that tip is not cut back in winter, the tree will come out of dormancy in the spring and will not shoot up because the brake is already in place. By reserving winter or dormant pruning for structural cuts and the removal of dead, diseased, and damaged wood and by making heading back cuts in the summer, you prevent unnecessary vegetative growth and allow shoots to mature into fruiting wood."

Back at the Sunriver Community & Allotment Gardens, Clay was feeling good about the young trees now growing "in the middle of a subdivision." He was less concerned about grafting every single heritage tree in Sooke than he was about preserving different varieties. And despite the irony of planting new trees, which are years away from producing fruit, next door to 140-year-old trees that produce 600 pounds of apples, Clay can at least say that he is voluntarily helping preserve history and genetic diversity. "It's what makes me sleep at night. I know I've done my part as much as I can."

So what are Clay's favourite heritage apple varieties? His top five include Vanderpool Red, Fameuse (as the first apple known in Canada), Spartan (a good dual-purpose eating apple), Knobby Russet ("the ugliest known apple variety"—although Bob Weeden says his Dumelow is definitely a contender), and any variety that has Cox's Orange Pippin in its ancestry (including Holstein). This, he notes with a smile, adds about twenty varieties to his top "five."

LEMON PIPPIN
ENGLAND,
PRE-1700

Taste and appearance: Small- to medium-sized, often oval in shape, with greenish-white, crisp, acid flesh. The skin is pale yellow, tinged with green, changing to a lemon yellow as it matures.

Use: Eating fresh, cooking, and drying.

History: It is uncertain when the Lemon Pippin was first discovered. One source notes that the eighteenth-century writer of *The Modern Husbandman*, William Ellis, described the Lemon Pippin in 1744 as "esteemed so good an apple for all uses, that many plant this tree preferable to all others."

Growing and harvesting: Picked in October, it keeps until May.

PHOTO: CLAY WHITNEY

**VANDERPOOL
RED**
OREGON, 1903

Taste and appearance: Small- to medium-sized, conical in shape, with bright red skin and white, crisp, juicy flesh. Taste is slightly acidic and sweet.

Use: Eating fresh.

History: Originated in Benton County, Oregon, in 1903.

Growing and harvesting: Picked in mid- to late October, but the flavour improves with age and is best around Christmas time. It keeps in good condition until May.

BRAMLEY'S SEEDLING
ENGLAND, 1809

PHOTO: DERRICK LUNDY

Taste and appearance: A large, flattened, smooth, bright green apple, with yellow-white, firm juicy flesh and a tart, sharp, and acidic taste.

Use: Cooking.

History: Discovered as a chance seedling in 1809 in Nottinghamshire, England. The original tree was still alive as late as 1988.

Growing and harvesting: The tree is vigorous and grows well on the West Coast. Picked in mid-October, the fruit is "in season" from November to February.

Other: Ranks as one of the world's great culinary apples. Says one source, "Many cooks reach automatically for the trusty Bramley . . . Its key feature is the very high level of acidity, and the excellent strong apple flavour that [it] lends to any apple dish." [7]

PHOTO COURTESY KEEPERS NURSERY, FRUITTREE.CO.UK

ALEXANDER
UKRAINE, 1700

Taste and appearance: Large, round-conical fruit, yellow with red-flushed skin, and sweet, juicy, cream-coloured flesh.

Use: Eating fresh and cooking.

History: This old Russian dual-purpose apple originated in Ukraine *circa* 1700 and became popular in North America after 1817.

Growing and harvesting: Picked in mid-September and keeps well for a few months.

Other: Probably the parent of Wolf River.

Apple Brownies

3 large apples (any combination of varieties works)

½ cup softened butter

1 cup sugar

1 egg

½ cup walnuts (optional)

1 cup flour

½ tsp baking powder

½ tsp baking soda

1 tsp cinnamon

Peel, core, and thinly slice apples. Mix butter, sugar, and egg. Stir in apples. Mix together dry ingredients and stir into the butter mixture. Add a few drops of water to make the mixture spreadable. Bake in a buttered, 9-inch-square pan at 350 degrees for around 40 minutes. (Go short on the time—never overcook brownies.) Sprinkle top with cinnamon.

Recipe from the book *An Apple a Day*, which features hundreds of apple recipes by Salt Spring apple lover Mary Mollet[8]

Quinoa and Apple Salad

1 cup white quinoa

1 tsp honey

1 Tbsp finely chopped shallot

1 tsp cardamom

¼ tsp coarse salt

2 Tbsp fresh lemon juice

freshly ground pepper, to taste

2 Tbsp extra-virgin olive oil

2 Tbsp dried currants

1 apple, cut into ⅛-inch-thick wedges

¼ cup toasted slivered almonds

¼ cup loosely packed fresh mint leaves, coarsely
    chopped, plus more for garnish

Prepare quinoa as per package instructions. Fluff quinoa with a fork; let cool.
    Whisk together honey, shallot, cardamon, salt, and lemon juice in a large
bowl. Season with pepper. Whisking constantly, pour in oil in a slow, steady
stream; whisk until dressing is emulsified. Add quinoa, currants, apple, mint,
and nuts; toss well. Garnish with mint.

Recipe provided by Steve Glavicich, chef/owner, Braizen Food Truck, Calgary

CHAPTER SEVEN

"Comfort Me with Apples"

Dr. Bob Weeden checks
out the many varieties
of apples he grows, as
seen here in his cold
room at Whims Farm on
Salt Spring Island.

Dr. Bob Weeden with one of his Red Gravenstein trees. Fifteen of the original apple trees, planted around 1924, still exist and produce apples at Whims Farm.

Apple Facts

Apples appear several times in the Bible, including this verse from the Song of Solomon (2:5): "Stay me with flagons, comfort me with apples: for I am sick of love" (King James version). The World English Bible translation is: "Strengthen me with raisins, refresh me with apples; for I am faint with love."

Apples are used in many desserts like apple pie, apple crumble, apple crisp, and apple cake, and spreads such as apple butter and apple jelly. They can be baked, stewed, dried, or covered in toffee. Puréed apples make applesauce.

A toffee apple is a traditional confection made by coating an apple in hot toffee and allowing it to cool. Similar treats are candy apples, coated in crystallized sugar syrup, and caramel apples, coated with caramel.

Apples are eaten with honey at the Jewish New Year of Rosh Hashanah to symbolize a sweet new year.

It was Friday night and at least two popular bands were playing at local pubs on Salt Spring Island—tough competition for a lecture on apple-growing by an eighty-four-year-old expert from Seattle. So as I drove to the Lions Hall, I thought, "How embarrassing if no one shows up." Even more concerning was the fact that this marked one of my first introductions to the world of apples and I was well aware of my lack of expertise. I hoped I wouldn't be put on the spot amid this cozy group of what couldn't amount to more than ten people, and be forced to reveal my lack of apple knowledge.

The first clue that apples were bigger than popular bands occurred as I attempted to find parking in the crowded lot outside the hall. Then I stepped through the door and faced some fifty people, rows and rows of chairs, desserts, coffee, tea, and a buzz of apple-growing chatter.

As the evening unfolded, I realized there exists a boundless wealth of information about apples, and people like Dr. Bob Norton have dedicated lifetimes to mining it. Dr. Norton has been growing apples since he was eighteen. He ran a tree-fruit testing program at Washington State University for over 30 years, and was responsible for testing 325 apple varieties, including Washington's first French and English cider varieties. Dr. Norton travelled the world, looking at apple cultivars for this program, both new and heritage varieties, aiming to find apples suited to growing in the Pacific Northwest. He also compiled a list of heritage apples cultivated in this part of the world prior to 1917. He's an in-demand commodity at apple-identification tables—a guru to people like Clay Whitney—and is known for introducing the "firm, crisp and flavourful" Jonagold to commercial orchardists in the Puget Sound region. (Jonagold, a cross between Golden Delicious and Jonathan, was originally developed at Geneva, New York, in 1943.)

With a background in both chemistry and biology, Dr. Norton was able discuss in detail solutions to various apple-growing problems, and answer the many questions put to him at the meeting about everything from mason bees to chip-budding. I became more and more dazed at the depth of information.

As my apple research continued I met a lot of apple people, some knowledgeable but incomprehensible, and others, like orchardist Dr. Bob Weeden, who were both informative and blessedly understandable. Like Dr. Norton, Bob has learned through trial and error (but at a grassroots level) about the "dos and don'ts" of apple growing. His love of history and the poetic word add another level of intrigue to his story. And through him, I discovered Janaki Larson, a forward-thinking grocery store owner, who sells Bob's heritage apples in Vancouver. As I stood back from my research, I liked the way these three people, though entirely different, were connected at the core by a love of heritage apples. Each, in his or her way, was contributing to the knowledge, understanding, and perpetuation of this living link to our past.

Bob's background is based in academia. A biology professor at the University of Alaska, Bob also focussed on environmental ethics and politics, and was one of the first to recognize the significance of environmental law as it was emerging. Over the years, he has evolved from scientist to naturalist and now, from professor to orchardist. He is the cultivator of 120 different varieties of old apples. Some of these are unnamed and unknown, grafted, for example from trees found on the roadside near his seventeen-acre farm on Salt Spring Island.

"There is a tree on Beddis Road that I watched for a couple of years," he told me. "It was a big tall tree with big green apples."

He snipped a twig, grafted it, and now has a tree of the same variety growing in his orchard. It has no name, but it's a good apple—one of the few chance seedlings that tastes good.

"It's such a long shot, though. To go through the whole grafting process and wait for fruit on a tree that you don't know anything about."

Bob's decision to move from Alaska to Salt Spring and grow heritage apples—after having not "even seen an apple tree in thirty years"—marked the intertwining of several factors. His life in Alaska was busy as he worked and raised a family with his wife, Judy Weeden (now an accomplished potter on Salt Spring), and it became important as he looked toward retirement that he give back to the community.

"A lot of what I did was head work," he said. "I was working, going to meetings, writing and lobbying—I didn't do anything for the community. I was tired of working with my mouth. My dad was a hands-on person and I wanted to dig and plant and see what grows. I was in the mood to work with my hands. In teaching, you

One of Bob Weeden's late-ripening heritage apple varieties—Seaford —was taken from a seedling from a now-defunct nursery on Bowen Island.

never really see the fruits of your labour. Working on a farm, the results are right there in front of you."

As a twelve-year-old boy living in New England, Bob earned money by picking up windfall apples at a commercial orchard. He enjoyed the money—a boy in the Depression didn't get to see many coins—but also, he remembers the apples: "The smell of McIntosh was very much a part of my youth. People didn't plough or scorch the earth to keep the bugs down in those days, so sitting under a tree, eating a windfall apple, was very different than today. You'd be sitting in a field of flowers and grass," he recalls. "I always had an idea that someday I would like to grow those trees—the McIntosh,

Wolf Rivers, Northern Spies, and Golden Russets of my New England youth."

Intervening school and work years didn't dim the nostalgia, so when he pondered retirement, "a yearning to grow apples rose to the surface." He and Judy first visited Salt Spring on a bicycle trip in 1987. They returned the following year and bought Whims Farm, a parcel of land that still had trees planted by the owner's father in 1924. Fifteen original trees still exist, including Kings, Gravensteins, Golden Russets, a Red Astrachan (Russia, 1800s), a Baldwin, a Rhode Island (one of the oldest known American apples, 1650s), and an old crabapple, described as "sweet for a crab, but still pretty tart and small."

A few years later, in the early 1990s, several other farmers started thinking about adding heritage apples to their endeavours, and Bob joined the process. One of the people leading the heritage apple revival at the time was Renée Poisson, who ran Tsolum River Fruit Trees in Merville on Vancouver Island. Renée had over three hundred types of heritage apples and pears and "played a big role in carrying on the traditional varieties," says Jane Lighthall, of Denman Island Heritage Apple Trees, a small business that propagates and sells over ninety types of heritage trees on Denman, a small island just east of Vancouver Island. The Denman business started in part because of the niche created when Renée retired, but the boon to Salt Spring growers was even bigger as they split her collection of three hundred trees, and moved them, with the help of a grant from the Canadian Organic Growers Association.

"I got close to 100 trees," Bob said. "Some died over the next decade but I replaced them with scions from other nurseries. My orchard, now 18 years or so in the making, has 120 varieties of apples and this year bore close to 4 tons of fruit."

At one point Bob had 170 varieties among his 200 trees, but he lost several dozen to canker. He has stopped replacing them because the orchard is a lot of work—at 78, Bob grows, harvests, and sells all his apples on his own. "The curve of production is going up," he laughed, "as my level of energy is going down."

Luckily, the trees ripen at different times, with the first of the Gravensteins ready in mid-August, and the bulk of the harvest plucked throughout September and the first two weeks of October. Finally come the late-ripening varieties, like Seaford (from a seedling from a now-defunct nursery on Bowen Island), which is not ready to eat until after Christmas. Some, like the Yellow Transparent (Russia, before 1870) "come so quickly with the sugar turning to starch, they have to be picked in two days."

Typically, one quarter to one third of Bob's crop is pressed into apple juice (yielding about 150 two-litre jugs), which can then be frozen, stored, and sold all year round.

"The secret of good juice," Bob said, "is mixing different varieties of apples." However, he always ensures he has Golden Russets in the batch because "they have a nice deep colour and add body, so the juice won't be thin."

Bob chose medium-sized, semi-dwarf rootstock for his most of his trees, but put the Gravensteins on standard (larger) size, because they are genetically more vigorous. His Gravensteins are about twenty feet tall while his other trees range from fourteen to sixteen feet. Like Harry Burton, Bob is not a fan of dwarfing rootstock, because the resulting trees have poorer root development, and need more water and fertilizer.

Dwarfs have advantages, he said, especially to commercial growers, as they mature faster, taking two to three years to bear fruit compared to six to seven years. "You can pick them without a ladder, but so can deer. Commercial orchards are willing to deal with water, fencing, and fertilizing for the convenience. The savings in labour are huge. And because the fashions in apples change rapidly, they don't want to plant a tree that will be producing for fifty years—it's better to tear out twelve-year-old trees that already have produced crops for nine or ten."

Bob's interest in heritage apples, combined with his academic background and membership in the BC Fruit Testers Association, prompted him to travel to smaller, sparsely populated islands around Salt Spring—Parker, Wallace, Portland—and reproduce via grafting some of the heritage apple trees in old, often overgrown orchards there. He also visited fifty-five farms on Salt Spring and created a list of apples currently in production and, as well, researched varieties grown before 1920, compiling another list.

Bob is also a writer and has published a beautifully poetic, first-person account of apple history, *Tian Shan and Chuan*.[1]

"I float through forty million years of yesterdays . . . Close by is an apple, its dangling, bitter fruit small and undistinguished. The forest

A youngster enjoys a little fun amid a huge pile of apples during one of Salt Spring's famous apple festivals.

PHOTO: DERRICK LUNDY

is east of Eden, beyond Nod, on a sprawling, fertile plain which later will be called China."

From here, he follows "the way of the apple" as it spreads across the world, propagating in the wild and then through the "transforming" invention of grafting. Finally, he arrives at his orchard on Chuan (the Native name for Salt Spring), and eloquently traces the history of some of his own apples.

"I am, I think, a man imitating an orchardist. I hold a basket, the basket holds the harvest of ten million years. In it is a decorative branchlet of Lady apples, cheerfully red and green, exact copy of a fruit found or brought by the westering soldiers of the Roman Empire, saved from barbarians later by Cistercean monks . . . There are Calvilles, lumpy and freckled but lively to the taste. Claude Monet painted this apple; Richard Harris, fruiterer in East Kent, brought supplies of this kitchen queen for the enjoyment of Henry VIII during beheadings.

"I have picked three Catheads in hope that they will become a well-cinnamoned pie. The Cathead was grown in England when Acadians and Pilgrims readied for sailings to the New World. Given its productivity and rebellious shapes, twigs of the Cathead might well have been among the cargo of the tiny ships . . ."

Bob's lyrical essay is based, in part, on Dr. Barrie Juniper and David Mabberley's *The Story of the Apple.*

Bob's interest in the historical anecdotes of his apples is compelling, but he's also part of the future of heritage apples, this year sending three hundred pounds—including Kandil Sinap, Golden Russet,

King, Gravenstein, and Smokehouse (Pennsylvania, 1837)—to a specialty food store in Vancouver. Le Marché St. George, run by sisters Janaki Larsen and Klee Larsen-Crawford as well as Janaki's partner, Pascal Roy, is a niche grocery store, riding the wave of the Slow Food Movement, located in a residential neighbourhood between Main and Fraser Streets.

"Most neighbourhood stores sell cigarettes and lotto, cards, and junk food," Janaki explained. "We thought it would be more interesting to revisit the classic concept of a general store. We stock everything from organic farm eggs, milk, artisan cheeses, sausage, to flour and spices. For produce we are trying only to carry local and in season."

Janaki spent some of her teen years living near an old apple orchard and she "loved nothing more than walking out my front door, across the field, to pick apples right off the trees. I loved their unrivalled flavours, their perfect textures, and imperfect skins."

She said it's difficult to find heritage apples in the city—even organic distributors aim to sell apples chosen for their consistent shapes, perfect skins, and conformity. "I am more interested in *real* food, not in a trendy way, but in the way that I had access to these kinds of foods growing up. We carry as many heritage varieties of produce as we can."

Of Bob's apples, Janaki said, the Kings were by far the most popular. "The more conventional of the heritage apples sold first. People definitely stick with the things they are familiar with. I personally loved the Turkish ones."

The biggest challenge of selling the apples, she added, is that people still expect their "heritage, organic, local" apple to look perfect. But of the three hundred pounds they purchased from Bob in October, only ten pounds remained in December. "I thought they did very well. People were very excited to see them and try them out."

She sees Vancouver as a "food-savvy city," noting that consumers, and especially chefs, know their stuff and are always looking for something new.

"The movement for local food is very strong, but I still think the area of 'heirloom' produce is under-utilized and under-marketed. I think that will be the next wave of 'it' foods."

Some Heritage Apple Cooking Tips from Judy Weeden

Golden Russets make superb baked apples. A key ingredient in the cored apple, besides brown sugar or maple syrup, is mincemeat.

Cathead, Bramley's Seedling, and Gravensteins make good pies. Sugar to taste, then a dash of lemon. Overcooking makes apples mushier.

Several early-season apples are great for sauce. Yellow and Red Gravensteins are equal in texture and taste, but Red Gravenstein makes a pink sauce. So do Red Astrakhans, which are hardy and grow well here on the West Coast.

CALVILLE BLANC D'HIVER
FRANCE, 1500s

Taste and appearance: The ugly exterior of this misshapen, green apple belies a sublime interior. The flesh is tender, juicy, and rich-flavoured.

Use: Eating fresh, cooking, and juice. It holds its shape in cooking, but if cut into small pieces, it will dissolve into a rich, sharp-textured purée.

History: Originated in Normandy, France, around 1598, where it was grown by Louis XIII. In America, Thomas Jefferson is said to have grown it at Monticello.

Growing and harvesting: The trees do best in warm soil against a sunny wall or bank and need long, hot summers to mature. Picked in late October, the fruit keeps until January.

Other: It is said to have a higher vitamin C content than an orange.

CORTLAND
NEW YORK, 1915

Taste and appearance: A medium-large, flat-round, yellow-striped red apple. Flesh is crisp, white, and juicy, with a tart and tangy flavour.

Use: Eating fresh and cooking—especially salads as the slices are unusually slow to brown.

History: Released from the New York State Agricultural Experiment Station in Geneva in 1915.

Growing and harvesting: Can be prone to scab and canker. Picked in late September, it keeps until January (or May in controlled storage).

Other: A cross between Ben Davis and McIntosh. Widely grown in Quebec and Ontario, having proven well-suited to freezing temperatures.

PHOTO: CLAY WHITNEY

Taste and appearance: Cathead is one of the oldest apples known in England. The name comes from its alleged resemblance to a cat's head—the shape is unusually conical and can be ribbed.

Use: Primarily cooking. The flesh is juicy with a fair amount of acidity, but does not need much additional sugar when cooking.

History: Originated in the early 1600s, possibly from the Severn Valley in England. One source claims Cathead is the first known variety planted by early settlers in Virginia, possibly as early as 1620.

Growing and harvesting: Easy to grow but slow to start bearing fruit. Ripens in mid- to late season.

KANDIL SINAP
TURKEY,
EARLY 1800s

Taste and appearance: Tall, slender, cone-shaped apple, with such a narrow base it hardly stands upright. Creamy, yellow, porcelain-like skin with red blush. Crisp, juicy, fine-textured flesh, excellent flavour.

Use: Eating fresh and cooking (purée, applesauce, apple butter).

History: From Turkey (but a few sources say Russia) in the early 1800s. It was sold commercially in Great Britain by the 1860s.

Growing and harvesting: Dwarfish tree grows in narrow, pyramidal form. Picked in mid-October and keeps until February.

Other: Because of its excellent flavour and unusual shape, it is becoming popular with apple connoisseurs in North America.

Apple-Onion Tarts

1 Tbsp extra-virgin olive oil

1 Tbsp unsalted butter

3 medium apples, peeled, cored, halved, and
  sliced ¼-inch thick

8 medium yellow onions, halved and thinly sliced

3 Tbsp cider vinegar

½ tsp coarse salt

2 Tbsp chopped fresh rosemary

1 cup (or 2–3 oz) coarsely grated manchego
  cheese

freshly ground pepper, to taste

6 3-inch frozen tart shells

Heat oil and butter in a large skillet over medium-high heat. Add apples
and onions, and cook until golden brown, about 15 minutes. Cover, reduce
heat to low, and cook until very soft and caramelized, about 35 minutes.
Add vinegar and salt, and cook 5 minutes. Let cool. On a baking sheet,
lay out tart shells, sprinkle with chopped rosemary. Using a spoon, add
3 tablespoons apple-onion purée to fill each tart shell. Sprinkle each with
2 tablespoons cheese. Season with pepper. Preheat oven to 350 degrees.
Bake until edges are golden brown, 20 to 25 minutes. Serve tarts warm or
at room temperature.

Recipe provided by Steve Glavicich, chef/owner, Braizen Food Truck, Calgary

CHAPTER EIGHT
Deep Roots

Bev Sidnick, of Hoodoo Ranch at Spences Bridge, BC, shows an apple tree where a bear has reached up, pulled down a branch, and started eating the apples. Bears were a huge problem at the ranch until Bev and her husband Bob Howard obtained three large dogs.

Hoodoo Ranch gets its name from the huge hoodoos that rise above it (partially seen here on the left). The ranch, one of the few in BC that grows heritage apples, is a spot of greenery amid the desert-like landscape.

Apple Facts

Facts gathered by BC Agriculture[1]

British Columbians consume 25 per cent of the apples grown in BC. That's about one hundred apples per person per year.

In nature, a protective wax shield covers most plants, flowers, and fruits. While an apple is still on the tree it develops this coating of plant wax, which slows dehydration.

Apples are the second most popular fruit sold in supermarkets, ranking next to bananas.

In Norse mythology, King Rerir prayed to have children. Hearing his plea, the Goddess Frigg sent him an apple of fertility, by way of a crow, which dropped it in his lap. The queen ate the apple, resulting in a six-year pregnancy and the eventual birth of their hero son, Volsung.

In February 1884, Jessie Ann Smith prepared to leave her home in Scotland with her new husband, fruit-growing expert John Smith. They were headed to Spences Bridge, part of British Columbia's dry belt, where today, standing amid the cactus and sagebrush on the ridge above Hoodoo Ranch, you can still see the line in the hillside where the Cariboo wagon trail once carried gold-seeking prospectors.

"I would like you to take these with you," said Jessie Ann's father, pointing to several little apple trees he had growing in his garden. "It is a new variety of apple called Grimes Golden. John might like to try it in the orchard he is developing."

So begins Jessie Ann Smith's adventure-filled memoir, *Widow Smith of Spences Bridge.*[2] Her story covers the gamut of perilous exploits associated with pioneering a new land, including confrontations with rattlesnakes, bears, and cougars; brutally cold winters; a herd of cattle drowning on thin ice; and John Smith surviving a mining accident, in which he was buried up to his neck in a slide. But it is also the story of Jessie Ann's love affair with apples, especially the Grimes Golden, for which she became famous. Even today, in Spences Bridge, you can see one of her beloved trees—unmarked and unadorned, but still a landmark of living history.

Amid all their adventures, the Smiths had created a substantial orchard of over 3,000 trees by the time John died in 1905. Now the "Widow Smith," Jessie Ann, and her children set about maintaining the orchard and selling the fruit which, at its peak, amounted to 12 tons of apples—contained in 20 railway box cars (each holding

630 boxes) and sent to Vancouver, Calgary, and beyond—along with cherries, apricots, peaches, plums, and pears. But it was through fruit exhibitions, which the BC government began using as a means to promote exportation, that Jessie Ann's fame grew. When individual growers were asked to submit exhibits to the Royal Horticultural Society (RHS) in London for the Colonial Fruit Show, Jessie Ann sent a collection of apples and won a silver medal. She began exhibiting both overseas and in the United States and won numerous medals, ribbons, and even cups.

"In 1909," she wrote, "I sent another exhibit of Grimes Golden and other apples to the show in London by the RHS. King Edward VII visited this exhibition and asked to see Widow Smith's Grimes Golden apples. At first my apples could not be found. The officials at the show were nearly frantic. They tried to show the King other apples belonging to the Smiths in Devon, others grown by Smiths in Kent, by Smiths from everywhere, but the King was not satisfied. 'The apples which I have come to see are those of the Widow Smith of Spences Bridge, BC,' the King persisted. At last, to everyone's relief, my apples were located."

This charming piece of BC apple history sets the backdrop to Hoodoo Ranch, which is one of very few orchards growing heritage apples in BC's interior. Owned by Bev Sidnick and Bob Howard, the ranch is a lush slab of greenery amid a desert-like landscape, stretching alongside the Nicola River and below a towering wall of hoodoo formations. The couple has close to a thousand apple trees, including a collection of more than two dozen heritage varieties.

Some are so rare they can't be found anywhere else in the region. Their more unusual trees include Ananas Reinette (Netherlands, 1821), Ashmead's Kernel (England, 1700), Egremont Russett (England, 1880), Orenco (USA, 1867), and, of course, Grimes Golden (USA, 1804).

I arrived at Hoodoo Ranch on a sunny fall day in October, basking in the beauty of the landscape around Spences Bridge and thrilled by the scenery on the drive up Fraser Canyon, along the Trans-Canada Highway from the Lower Mainland. My frequent trips through BC and into Alberta have typically occurred via the Coquihalla, a fast-paced highway constructed in the 1980s. While that trip is beautiful too, I realized I hadn't taken the original route since the Coquihalla was built. I let my mind toy with this as a metaphor for new and heritage apples. The route I usually take is faster and more convenient, but by travelling it, I'd forgotten the intense joy of the original highway.

With this in mind, I toured Hoodoo Ranch, sampling some choice, juicy fruit and admiring the pristine appearance of the trees and their apples. Bev showed me one tree that had been damaged by a bear, a huge limb broken and lying perpendicular to the trunk.

"The bears reach up and pull down a branch. Then they just sit there and strip the apples off it."

The addition of three big dogs to the ranch helped solve the bear problem.

Bev and Bob purchased the 129-acre Hoodoo Ranch in 2008. Bev is a retired special education teacher and long-time avid

Bev Sidnick believes this old Golden Grimes tree near the highway at Spences Bridge once belonged to pioneer Jessie Ann Smith, who became famous for her apples.

organic gardener, while Bob is manager for the area's highways contractor. Both spent years horse ranching. Before buying the ranch, they purchased and restored the historic James Alexander Teit house in Spences Bridge. Although the ranch was overgrown with cactus and sagebrush when it came on the market, it presented them with a new challenge—one they have embraced with transformative results. In buying Hoodoo, Bev said, they'd met some of their goals, like living a simpler life, growing food organically, becoming more self-sufficient, leaving a smaller environmental footprint and contributing locally, adhering to slow and local food movements.

"After three years of searching for an affordable property, we had almost lost hope when the property came available just five kilometres from where we were living," Bev said. "As luck would have it, it was a certified organic farm complete with almost a thousand trees—cherries, peaches, plums, apricots, pears, and apples. Included in the orchard were apple trees that I hadn't heard of since my youth."

Farming apple trees at Hoodoo is full circle—with a twist—for Bev, who grew up on an apple orchard near Vernon, BC. Coldstream Ranch, a large commercial operation, specialized in Spartan, McIntosh, and Delicious.

"My first paying job was picking apples back in the days when the use of pesticides went unquestioned," Bev recalled. "Large tractors with huge tanks full of toxic chemicals slowly proceeded up and down the rows of trees, spraying billowing clouds of these chemicals

OPPOSITE:
Crates of apples line the orchard at Heart Achers Farm as the fruit is harvested. Heart Achers has half of its thirty-five acres planted in apple trees, with six hundred trees per acre, producing two hundred thousand pounds of apples.

into the air, ensuring it would blanket the entire area. This spraying was done numerous times during the growing season. As children we would run and play in the orchards, eating apples in their every stage from small unripe green apples to fully ripe apples ready for harvest. I spent months every year handling and consuming the apples, not realizing the potential health risks."

Hoodoo's operation is tiny in comparison, producing about fifteen thousand pounds of apples, but the couple is able to work the entire farm with a small amount of assistance from friends and family. They sell apples to Discovery Organics in Vancouver, Footprints Harvest CSA in Merritt, at farmers' markets, via farm gate sales, and through a "pick your own" program. They also give back to the community by donating apples to several organizations, including the Elizabeth Fry Society in Ashcroft, Heskw'En'Scutxe Health Services Society, and Hidden Mountain Drummers (Lytton First Nation).

Bev picked Cox's Orange Pippin as one of her top eating apples, but said, "I suppose my very favourite is Gravenstein. [It's] acceptable for eating, but unsurpassed for applesauce and pies. Many of my mature customers request this apple as they also remember it from their childhoods."

Hoodoo introduces customers to the different flavours of the apples by free taste testing on the farm, at farmers' markets, and by donation.

"Often our u-pick customers who arrived on the farm for a specific variety end up tasting, preferring, picking, and leaving with a

Ron Schneider of Heart Achers Farm in Cawston, BC, another of the few orchards in BC's interior growing heritage apples instead of more commercially viable types.

different variety. I always tell my tour guests that we operate an 'all you can eat' orchard, encouraging them to try everything."

Hoodoo's farm tours teach visitors about organic farming and natural pest control among other things. The ranch also offers riverside camping and other activities, with lots of information on its website.[3]

After leaving Spences Bridge, I drove east across to Merritt, Kamloops, Revelstoke, and then down through the Okanagan apple-growing regions (past Coldstream), Kelowna, Penticton, and all the way down to Cawston, near Osoyoos on the province's southern border. Here I stopped at Heart Achers Farm—the only orchard in the province that the British Columbia Fruit Growers Association was actually able to name as specializing in heritage apples. The association knew of Heart Achers' Ron Schneider and Andrea Turner because the orchard's Cox's Orange Pippins won the Canadian National Award for Heritage Varieties at the 2008 Royal Agricultural Winter Fair. The two have been farming their organic heritage apple orchard for more than twenty-five years. They are also part owners of Direct Organics Plus packing house and distribution centre for organic growers in Cawston.

By the time I visited this 35-acre orchard in the heart of BC's organic-growing country, they had already harvested their Gravenstein, Hyslop, Fameuse, Bulmer, and Porters, most of which (150,000 pounds) went to Sea Cider in Victoria. They were set to pick Cox, Calville Blanc, Winesap, and Newton Pippins. In total, they have 15 varieties of commercial heritage apples, with their Newton Pippins and Winesaps being the most popular on the fresh market. The vast majority goes to cider makers, and some goes to a nearby winery called Rustic Roots. (Run by the Harker family, Rustic Roots Winery makes several fruit wines, including one that uses Fameuse from a 110-year-old tree that still produces apples on their fifth-generation family farm.)

Heart Achers took a risk, planting heritage apples rather than commercially proven varieties. However, the risk appears to be paying off, as interest increases, and Heart Achers may turn out to be one of those forward-thinking businesses now on the edge of something that's about to boom. (Similarly, years ago, they were among the first to jump on the organics train.) Most recent interest in Heart Achers apples, said Ron, has come from cider makers and cider-apple growers, another group of entrepreneurs getting into the basement of a burgeoning business. But he's also had calls from elsewhere, like a scientist from the University of Saskatchewan's fruit program, wanting to test heritage apple trees suitable for growing on the prairies.

As Ron pointed out, most orchards go with "safe" apples because sales have to be high to make land prices (between seventy-five thousand and one hundred thousand dollars for a bare acre in Cawston in 2010) worthwhile. Irrigation, labour, and equipment are added costs, and for Ron, whose family has farmed the land in Cawston for several generations, farming may be in his blood, but he's not getting rich from it. ("Why do you call it Heart Achers?" I asked. Said Ron, "If you've been farming long enough, you find out.")

Today, Heart Achers has half of its thirty-five acres planted in apples, with six hundred trees per acre, producing two hundred thousand pounds of apples. At age sixty, Ron said he's "winding down" but his sons are gearing up and taking on more and more of the work. After a tour of the orchard and the packing-and-distribution house, I got back on the highway, a sample of Ron's generosity packed into the back of the car. It included some of the best-tasting apples, pears, and plums I have ever experienced, a bottle of Sea Cider's cider, and an absolutely divine-tasting sparkling apple wine from Rustic Roots Winery.

Back on the coast I found yet another piece of living apple history, this one at Brae Island Heritage Apple Orchard (located on Allard Crescent in Langley), where a group of volunteers rescued heritage apples trees planted in 1858 by the Royal Engineers of Fort Langley.

Arborist Bill Wilde, a member of the Derby Reach/Brae Island Park Association, which took on the project in 2008, said several years ago they realized "there were a number of quite old apple trees [in the park]—between eighty and one hundred years old—along the river bank for several acres. We took it upon ourselves to clean it up, which meant taking out a lot of blackberry bushes. We also worked on remediation and abatement of the old trees, cleaning up and pruning."

The trees included Northern Spy, Baldwin, Winter Banana, Wolf River, and Blue Pearmain, as well as several that remain unidentified.

"We got some oral history from one of the families that used to farm in the area. [One older gentleman] could remember living on the farm as a little boy with a black man living in a shack on the land. He could remember playing in the orchard. This gave us a real timeline."

They decided to take cuttings from the identified apple trees as well as one unidentified pear tree, and graft them onto medium-sized rootstock. From there it seemed a good idea to create a heritage orchard; however, the park is on a culturally sensitive Kwantlen First Nation archeological site.

"The cost and details of having an archeologist on site for the digging and planting of the trees proved prohibitive," Bill said, so he went "back to the drawing board" and resurrected a tree-planting method (called the low profile pot system), which he'd used previously in New Westminster and requires no digging at all. After a presenting the plan to the Kwantlen First Nation and their advisors, he received approval to go ahead with it.

Basically, small trees are placed on heavy black plastic sheets filled with soil and then surrounded by sheet metal, a metre in diameter. The trees' roots spread outwards in the soil within the metre-wide cylinder, and the trees themselves are kept vertical on wire lines like grapes.

"Then you tease the root mass up and out, and pull out the plastic. Now all the roots are exposed to the soil underneath. It's nice to have them in the location where they will reside, but we had to move them to a lawn area."

The trees were placed on mounds of soil and then back-filled with additional soil—no digging required. Planted in 2011, the trees will take three to four years to grow and then the wire will be removed.

"The plan is to continue to propagate the trees on a scaled-back basis," Bill said. "There's lots of enthusiasm for the project."

The park association is all volunteers, who work in conjunction with Greater Vancouver Regional Parks and take on projects such as the removal of invasive species, replanting trees, and the preservation of a bog in the park.

"In the context of the Slow Food Movement, this is locally produced food. We know where it comes from. It is so much better than eating these hyper-hybridized varieties made because they last longer

Apples and other organic fruit—some from Heart Achers—are on sale at a funky roadside stand in Cawston, BC.

and look good," said Bill. "I've always thought of trees as pieces of living heritage. Unlike other things, like monuments or buildings, this is something that was here, for example, when the railway went through. These trees were extremely important to those people who planted them."

This is a sentiment, I'm sure, with which all the Royal Engineers of Fort Langley, as well as Jessie Ann Smith of Spences Bridge, would undoubtedly agree.

WINESAP
UNKNOWN ORIGIN,
1700s

Taste and appearance: Medium-sized, conical, bright deep red over a yellow background with yellowish, crisp juicy flesh. Taste is sweet, sprightly, and aromatic.

Use: Eating fresh, cooking, juice, and cider. Primarily a culinary apple.

History: Was a major commercial variety in Virginia during the nineteenth century. Its origins are unknown but it probably dates back to the eighteenth century.

Growing and harvesting: The trees are vigorous but tender in the cold and are unusual in that they have pink blossoms instead of white. Ripens in late October and keeps until April.

Other: It was considered to be a good dual-purpose keeper until the 1950s, when better-flavoured, longer-keeping apples were developed.

PHOTO: SIERRA LUNDY

Taste and appearance: Medium-sized, oblong, and often flattened at the ends. The skin is somewhat rough, clear yellow with a slight red blush and fine russet dots. It has a rich, distinctive aromatic flavour, and bruises easily.

Use: Eating fresh, cooking, juice, and cider.

History: Originated as a chance seedling in West Virginia in the 1830s, and was widely planted during the early twentieth century. Its high sugar content was put to good use in brewing hard cider.

Growing and harvesting: It ripens in early October and keeps until February.

Other: Grimes is possibly a parent of the Golden Delicious. It is one of a relatively select group of apple varieties that are self-fertile.

BALDWIN
BOSTON, 1750

PHOTO COURTESY ORANGE PIPPIN LTD.

Taste and appearance: Large, red and green. The flavour is described as "sweet and unpretentious, crisp and pleasant."

Use: Eating fresh, cooking, and juice. Retains its shape when cooked and lends a rich, sweet flavour to apple pies.

History: Originated as a chance seedling in the mid-1700s. Initially called Woodpecker or Pecker, it was very popular in the US in the nineteenth and early twentieth centuries.

Growing and harvesting: Picked in late October or early November and keeps until February.

Other: Baldwin was one of the most important American commercial apples in the nineteenth century, being an excellent keeping apple with a fairly thick skin, and therefore able to withstand long-distance transportation.

PHOTO: DERRICK LUNDY

EGREMONT RUSSET
ENGLAND, 1872

Taste and appearance: Medium-sized, flat-round with golden skin, and orange flush and ochre russeting. The flesh is cream-coloured with an aromatic, nutty flavour.

Uses: Eating fresh and cooking.

History: First recorded in Somerset, England, in 1872.

Growing and harvesting: Picked in late September, it stores until December.

Spiced Crabapples

4 lb crabapples

2½ cups apple cider vinegar

2 cups water

4 cups sugar

1 Tbsp whole cloves

3 cinnamon sticks

1 tsp fresh ginger

Wash apples and leave stems on. Prick each apple in several places with a needle. Bring vinegar, water, and sugar to a boil. Add spices tied in a bag. Cook half of the crabapples in the syrup at a time, 2 minutes each. Pour syrup over apples and let stand overnight with spice bag. Pack apples into pint jars. Bring syrup to boil and pour over apples. Adjust lids and process in boiling water bath (212 degrees) for 30 minutes. These are great with roasted or grilled meats.

Recipe provided by Bev Sidnick of Hoodoo Ranch

Apple-Banana Bread

½ cup butter

2 eggs

½ cup sugar

½ cup brown sugar

3 bananas, mashed

1 medium apple, grated

⅛ tsp salt

1 tsp baking soda

2 cups flour

Glaze:

½ cup brown sugar

¼ cup butter

1½ tsp cinnamon

Blend butter, eggs, sugars, and fruit. Mix dry ingredients and add to wet. Bake for 45 minutes at 350 degrees in a greased and floured tin. For the glaze, combine sugar, butter, and cinnamon in a pot, heat slowly until sugar dissolves. Spread on top of bread at 45-minute mark, and cook 5 to 10 minutes more.

Recipe provided by Daphne Taylor, Salt Spring landscaper and caterer

CHAPTER NINE

The Cider House Rules

A cider-tasting event at Sea Cider in Victoria features this array of different tasting ciders. Many are made from apples grown by Ron Schneider at Heart Achers in Cawston.

Apple Facts

Cider or cyder is a fermented alco-holic beverage made from apples. Cider varies in alcohol content from 2 per cent to 8.5 per cent and higher in traditional English ciders. In some regions, such as Germany, cider may be called "apple wine." In North America, the term "hard cider" is sometimes used for the alcoholic beverage, while "cider" can also refer to a non-alcoholic apple juice.

It takes about thirty-six apples to create one gallon of apple cider.

It takes four to five years for a tree to start producing apples; a Northern Spy takes twelve years.

Fifty-six per cent of apples in Canada are sold fresh (at harvest or later), the remainder being processed into juice, sauce, pie filling, frozen slices, and other products.

"Here is the shortest possible instruction set for turning apples into cider," stated the website howtomakecider.com: "Get some apples; chop them up really small; press them; put the apple juice in a container; add yeast; let it ferment for a month or two; put the cider in clean bottles; wait for a couple of months; drink the cider."

As I read this, starting to feel thirsty, I thought, "Well, that sounds pretty easy." But then I read: "If you follow these instructions, you will produce cider, but probably not very nice cider. With brewing a drink that is both alcoholic and great tasting, the devil is in the detail."

The rest of the website was dedicated to de-deviling those details—an act that cider makers like Kristen Jordan, Rick Pipes, Janet Docherty, and numerous others around the world seem to have perfected. Cider is a fermented drink like wine, but made from apples instead of grapes. Like wine, cider can be still or sparkling.

Cider making is an age-old practice but it's difficult to establish just how early the first cider sippers emerged. One timeline I saw has villagers in Kent drinking a cider-like beverage made from apples when the Romans arrived in 55 BC. Apparently Charlemagne referred to it in the ninth century AD, and by 1066, cider consumption was widespread in Europe. Most sources seem to agree that following a decline in cider consumption in the twentieth century, it's now making a comeback.[1]

An increase in cider-drinking fits nicely with the growing popularity and demand for heritage apples. In fact, in some cases—such

OPPOSITE:
A container of apples is used—very appropriately—as a doorstop at Sea Cider in Victoria.

as Ron Schneider at Heart Achers—it's driving the demand. While cider can be made from any apple, the best cultivars are the bittersweet varieties grown specifically for cider making and not for apple eating. This means that cider makers must turn to heritage apples and not those made popular in recent years for their sweet taste and pretty appearance.

In Canada, cider is made in several provinces, much of it by small-scale craft cideries, but in Quebec, it's considered a traditional beverage. Quebec cider making has a colourful past, including a legislative omission in 1920 that rendered it illegal. The situation wasn't officially corrected until 1970, although cider continued to be produced (but not legally sold) in the interim. The revival of cider in Quebec is relatively recent because Quebec's alcohol-regulating body, the Régie des alcools, des courses et des jeux, did not start issuing permits for craft cider making until 1988. However, sources say that by 2008, some forty cider makers were producing more than a hundred types of apple-based alcohol.

One of these products, ice cider, is a Quebec innovation that uses the province's harsh climate to its advantage. Built on the idea of ice wine, ice cider is created by pressing apples that are naturally frozen on the trees during winter. The liqueur-like result is absolutely divine and it has taken the cider-drinking world by storm, winning numerous international awards since it hit the market in 1996.

I was introduced to ice cider in the summer of 2011, as Bruce and I wandered through a market in Quebec City. By about 9:00 AM, I found myself feeling pretty happy after sampling various ice ciders, which can have an alcohol content of up to 20 per cent. Later, I looked up the website of one of the ice ciders we bought—Le Pedneault—to see if heritage apples were used in its production. It turns out that Pedneault has a rich history, dating back to 1918 when the family first planted apple trees on ancestral lands, later launching a business in apple tree sales. It added a wine cellar and cider and vinegar factories in 1999, and now produces over twenty types of alcohol, mostly from apples, as well as vinegars and non-alcoholic products. In addition to growing some heritage apples varieties I'd seen elsewhere, like McIntosh and Fameuse, Pedneault grows Melba (Ottawa, 1898), Duchess of Oldenburg (Russia, 1818), Antonovka (Russia, 1800s), Lobo (Canada, 1898), and Wealthy (Minnesota, 1860).

A very different list makes up the heritage apple trees planted at Sea Cider, a small craft cidery on the Saanich Peninsula on Vancouver Island. Sea Cider has fifteen hundred trees, and grows over sixty varieties of organic heritage apples, selected for their "superior cider qualities"—which typically translates into "inferior eating qualities." Cider apples are usually too tart and astringent to eat, containing high levels of tannin, but they give cider a flavour that dessert apples don't have. Tannin, an important compound in traditional English-style cider apples, is defined as "any of various soluble astringent complex phenolic substances of plant origin . . ."[2] I personally find Kristen Jordan's definition a bit more palatable: "Tannin produces a tannic flavour, the kind of taste that black tea

A variety of apple offerings, created by Le Pedneault in Quebec, was being sold at a beautiful indoor market in Quebec City.

leaves in your mouth—big and dry. You want that for British-style cider. It gives the cider structure." Bittersweet apples have the most tannins or phenolics, and dessert apples the least.

In 2004, when Kristen bought the ten acres of pasture that now house Sea Cider, she and her then-husband, Bruce Jordan, decided to plant predominantly English bittersweet heritage apples like Dabinette, Kingston Black (also the name of their dog), Chisel Jersey, and Yarlington Mill.

"We thought when we planted the tannic-heavy apples that we'd go with tannic styles and make English-style cider, which has a wild, earthy flavour," Kristen recalled. "In fact, Sea Cider's Wild English cider is exactly that. Appreciated mostly by cider traditionalists, it's less approachable for North Americans, whose cider experience is often the syrupy-sweet pink liquid purchased in two-litre plastic bottles at the liquor store. Wild English is 'hardcore,'" said Kristen. "If you like stinky cheese, you'll like Wild English."

However, instead of making only English-style cider, they ultimately developed a mix of cider types. So in addition to their cider apple trees, they planted other heritage varieties—mostly of European and North American descent—like Fameuse, Gravenstein, Pomme Gris, Spartan, Summer Red, Winter Banana, and Wolf River. They also turned to apple suppliers like Ron Schneider.

"We thought we'd be self-sufficient but it proved better to work with others. It adds to the interest and complexity of the ciders," Kristen said. "Luckily, people like Ron had apples typically used for North American ciders, and others [nearby] are preserving and collecting heritage varieties that we don't see elsewhere."

It's exciting, she added, to use apples that have their own histories and stories, and to know the people who are growing and producing them. In addition to purchasing apples from Heart Achers, Sea Cider buys from orchards on the Gulf Islands and on Vancouver Island, as well as from LifeCycles, which, among other things, harvests and distributes fruit that would otherwise go to waste. LifeCycles supplies Sea Cider with all the apples (mostly King of Tompkins and Northern Spies) for its Kings and Spies cider.

For Kristen, knowing the stories behind the apples is part of the joy of making cider. Kings and Spies, which also uses whatever other apple varieties LifeCycles has available in a given year, is less dry than Wild English, and Kristen described it as a "fruit-forward,

Italian-style sparkling cider." When I tasted it, I was reminded of Prosecco, and Kristen agreed it's a cider that "sits in wine territory."

Kristen also likes the historical serendipity of the way Sea Cider makes Pommeau, a thicker drink, closer in taste and texture to port or sherry, which uses mainly Fameuse apples, many of which come from Heart Achers.

"Snow apples were first grown in Lower Canada, likely from a seedling brought over from Normandy. Now we're using those to make a Normandy-style cider," she said. "Having the heritage behind the apples gives the cider a special character."

In making the Normandy-style Pommeau, Sea Cider first slowly ferments the hand-pressed apples and then takes the fermented cider to a distillery, Victoria Spirits. The cider is distilled and then aged in oak barrels for about six months. The alcohol level sky-rockets to 90 per cent, and Sea Cider adds apple juice to bring it down to 19 per cent. (In France, it must be 17 per cent alcohol to be called *pommeau* but typically it is between 17 and 19 per cent.)

Kristen was eighteen when her father died and left her an apple orchard on Shuswap Lake in the BC interior. She had no interest in apples, but developed a fondness for traditional-style cider during two years at school in Wales. "So there I was drinking cider, and I owned an apple orchard," she recalled, noting that the Shuswap trees had long since disappeared into the forest, their apples handily feeding neighbourhood bears. (Ultimately, it wasn't worth the work and expense of turning it back into a working orchard.)

But all this was fermenting in her brain years later, when, as

Bruce Cameron and Kristen Jordan discuss all aspects of cider and cider making, during a cider-tasting event at Sea Cider.

the mother of two young children, she worked as an international development consultant on agriculture rehabilitation in Ethiopia, and found herself wanting to be home more often with her family. (She now finds it ironic that she's educated in environmental management and said, laughing, "It's one thing to throw around advice and another to be the one actually doing the agriculture.") Suddenly pining for an occupation that would keep her at home, she started thinking about cider, which she and her husband had been making for fun with apples from trees on their Victoria property.

"We made cider for our personal consumption, and basically apologized when we served it to friends."

In order to "up their game," they took courses in commercial cider making at Washington State University. They found Ron Schneider online and, using his apples in 2002, produced about four hundred litres of cider in their garage. That amount was upped to four thousand litres by adding space in Kristen's mother's garage, and now Sea Cider produces forty thousand litres of cider a year.

"Ron was ahead of his time planting the heritage cider apples; he went against the grain," said Kristen, noting that he and other growers of heritage apples have helped preserve and maintain varieties that would otherwise disappear.

Most of Sea Cider's cider has been developed through trial and error, including its most popular variety, Rumrunner, which uses mostly Winesaps and Winter Bananas, some from Heart Achers. ("When Winter Banana ferments out," Kristen said, "it really smells like bananas.") Rumrunner resulted from an experiment in aging cider in screech barrels from Newfoundland. Unfortunately, Newfoundland stopped using oak barrels for screech so Sea Cider turned to Heavenly Hills in Kentucky, which now supplies them with bourbon barrels. Sea Cider resaturates the barrels by rolling screech around in them and then uses them to age the Rumrunner cider. (This doesn't always work according to plan—one batch of Rumrunner came out at 26 per cent alcohol and they had to blend it back to 12.5 per cent.) But the result is a dark, semi-dry, sparkling cider with hints of "brown sugar and spice" and, of course, rum. This is the cider putting Sea Cider on the map, selling all over Canada and making forays into the United States.

As a North American-style cider, Pippins is considered Sea Cider's most approachable brand. More like white wine, it can be paired with anything and makes "cider skeptics less skeptical," said Kristen. This cider uses mostly Newton Pippins from Ron blended with island-grown apples such as Winter Banana and Sunset. Like the pioneer North American cider makers, Sea Cider adds cane sugar during the fermentation process to raise the alcohol content to 9.5 per cent. (The higher the alcohol level, the longer it keeps, an important factor for old-time cider makers.)

Another cidery on Vancouver Island, located about ninety minutes north of Sea Cider in the Cowichan Valley, is Merridale Estate Cidery, the first of its kind in BC. Here, owners Rick Pipes and Janet Docherty sell seven ciders, three fortified dessert wines, and a range of brandies and spirits, all made on site from apples grown in their own orchard as well as from other BC farms. About twelve years ago, Rick and Janet were looking for a challenging and interesting business they could run together.

"We found Merridale," said Rick. "At the time, it was a small producer making about 20,000 litres of cider per year and having less than 1,000 visitors. We now produce and sell about 130,000 litres of cider and host about 30,000 people per year."

In addition to making and selling the ciders, wines, and spirits, Merridale has a bistro-style restaurant with local, seasonal food, a brick-oven bakery (yes, they make apple pie), a farm store—which sells gourmet food and local art in addition to Merridale products—plus a spa, featuring specialty cider soaps, foot scrubs, and other spa products made from site-grown apples. It offers tours (including four different cider tours), tastings, and lunch packages, and hosts events such as weddings in the orchard and special dinners at the restaurant.

Rick said Merridale ciders are made from heritage apples that have been used in Southwest England and Northern France for centuries, including Tremletts Bitter, Yarlington Mill, Dabinette, Frequin Rouge, Locard Vert, Julienne, and Judain. Merridale grows all of these varieties in its orchard.

"We also have five other orchards growing the same varieties for us [taken] from cuttings off our trees. Three of the orchards are in the Cowichan Valley, one is in Kelowna, and one is in Keremeos. By spreading out the growing regions, we get fruit with different flavour profiles and we spread out the risk of disease and weather."

As at Sea Cider, each of Merridale's ciders is a "blend of apples aimed at a particular palate." Some, like the House, MerriBerri, and Traditional, are designed to be enjoyed with or without food. Others, such as Normandie and Somerset, pair best with food, while Scrumpy and Cyser are more traditional English blends. Merridale's cider flights involve tasting six ciders, designed to introduce patrons to the "rich range and different traditions of cider making in one sitting."

"[There's] English, French, Scotch—even a taste of the ancient Viking-style Cyser. This is a history lesson most enjoy taking," states the cidery website.[3] In addition to focussing on their growing business, Rick and Janet are strong advocates for preserving healthy farmland for the future.

"Growing heritage apples using sustainable practices and educating people about the need to plan for the future is very important to Janet and me," said Rick. "Sustainable farming is about avoiding the use of herbicides and pesticides, and developing a healthy environment for the bees and the rest of our neighbours in nature. We're proud of the way we farm and work very hard to educate our guests."

Luckily, cider makers like those at Sea Cider, Merridale, and Le Pedneault are producing a number of palate-pleasing varieties that will meet head-on the rekindled desire for cider. This means that cider lovers like me don't have to try to figure out website de-devilling instructions on how to make our own cider in the basements of our homes.

KING OF TOMPKINS COUNTY

NEW YORK, 1800s

PHOTO: DERRICK LUNDY

Taste and appearance: A yellow apple with red stripes and flush. The flesh is yellowish, crisp, juicy, and somewhat coarse. It appeals to those who like aromatic, rich-tasting, sweet-tart apples.

Use: Eating fresh and cooking.

History: It is thought to have come from near Washington, Warren County, New Jersey, and brought to Tompkins County, New York, by Jacob Wycoff in 1804, who called it King. It was renamed King of Tompkins County in about 1855.

Growing and harvesting: Picked in early to mid-October, it keeps three months. Prone to mildew and scab, it is now losing its popularity to newer varieties that are easier to grow, though it is still highly recommended for the West Coast.

Other: Known commonly as King.

PHOTO: DERRICK LUNDY

NORTHERN SPY
NEW YORK, 1840s

Taste and appearance: Large to very large, round-conical, and sometimes slightly ribbed. The skin is green with dull red streaks or flush. The flesh is yellowish, fine-grained, and firm, with a high vitamin C content. It has a rich, intense, and fruity flavour.

Use: Eating fresh, cooking, juice, and cider. Described as the "supreme apple for apple pies."

History: Said to grow near the underground railway used by slaves escaping to Canada—thus its name. It started gaining popularity in 1840.

Growing and harvesting: Picked in late October, it keeps four months.

Other: Because of its resistance to the woolly aphid, the tree is used for rootstock breeding programs.

YARLINGTON MILL
ENGLAND,
EARLY 1900s

PHOTO COURTESY ORANGE PIPPIN LTD.

Taste and appearance: Small, firm, yellow apple with a red flush. The taste is sweet to bittersweet.

Use: Cider.

History: Emerged in England in the early 1900s.

Growing and harvesting: Described as "hardy."

Other: A very popular English cider apple.

**KINGSTON
BLACK**
SOMERSETSHIRE,
ENGLAND, 1820

Taste and appearance: Small, dark red, irregularly shaped.

Use: Cider and juice. Referred to as "the most valuable cider apple," with one of the best-flavoured juices.

History: Believed to have originated in Somersetshire, England, in about 1820, and probably named after the village of Kingston St. Mary, near Taunton.

Growing and harvesting: Picked very late in the season, and not the easiest of varieties to grow. It is generally considered prone to disease.

French Lentils in Cider

2 cups French lentils (green lentils)

3 cups chicken stock

2 cups dry cider

2 bay leaves

2 carrots, in ¼-inch dice

1 small red onion, minced

10 strips bacon, fried crispy and crumbled,
    or 8 oz hard sausage, cut up

Vinaigrette:

2 garlic cloves, minced

2 Tbsp Dijon mustard

3 Tbsp red wine vinegar

⅔ cup olive oil

1 tsp dried thyme

1 cup minced fresh parsley

coarse salt and freshly ground black pepper, to taste

Put lentils in pot and cover with stock and cider. Add bay leaves. Bring to a boil, then simmer 20 minutes. Add carrots and onions and cook until lentils are tender (10 to 15 minutes more). Drain lentil mixture (if necessary) but keep enough liquid to keep lentils moist. Discard bay leaves. Add bacon. Whisk garlic, mustard, vinegar, and oil together and pour over warm lentils. Add thyme and parsley and season with salt and pepper. Serve slightly warm or at room temperature.

Recipe provided by Kristen Jordan, owner of Sea Cider

Hot Buttered Rumrunner

3 cups (one bottle) Rumrunner cider

1 cinnamon stick

¼ tsp ground nutmeg

¼ tsp ground allspice

2 whole cloves

3 Tbsp brown sugar

1 Tbsp unsalted butter

Heat cider on low heat until it is hot but not boiling, then add spices, stir, and allow to steep for at least 30 minutes (best if allowed to steep for several hours). Add butter, stir to melt, serve in heatproof mugs. Makes 4 servings.

Recipe provided by Kristen Jordan, owner of Sea Cider

CHAPTER TEN

The Apple Doesn't Fall Far from the Tree

A worker at Heart Achers Farm is seen picking Spartan apples in mid-October. Although an "old" apple, Spartan is not considered "heritage" as it was developed at the Summerland Research Station in BC in 1936. It is a cross between a Newton and a McIntosh.

Apple Facts
Some famous apple quotes:

"Even if I knew that tomorrow the world would go to pieces, I would still plant my apple tree."
—Martin Luther

"Surely the apple is the noblest of fruits." —Henry David Thoreau

"Why do we need so many kinds of apples? Because there are so many folks. A person has a right to gratify his legitimate taste. If he wants twenty or forty kinds of apples for his personal use . . . he should be accorded the privilege. There is merit in variety itself. It provides more contact with life, and leads away from uniformity and monotony." —Liberty Hyde Bailey

"When the apple is ripe it will fall." —Irish proverb

In the summer of 2011, Bruce and I flew to Moncton, New Brunswick, to pick up my rustic 1978 Volkswagen van, driven from British Columbia the year before by my daughter Danica. As we planned the three-week, coast-to-coast trip, I blithely jotted down the locations of several heritage apple orchards, thinking we could drop by, take some tours, snap some photos. Once we were on the road, reality hit. First, Canada is really, really big. Second, a lot of apple-growing occurs in Quebec, and my high school French isn't interview-adequate. But most of all, it came down to the van, which, on a good day with a tailwind, reaches a maximum speed of ninety kilometres per hour. Drivers on fast freeways hated us, and almost immediately, we changed our route to secondary highways. Ultimately, this made the journey even better as we got a real flavour for small-town and rural Canada. But it also meant that a "side trip" to Nova Scotia's apple-growing region, the Annapolis Valley, was no longer "quick" or even possible.

One of the NS orchards on my list, Noggins Corner Farm, seemed impossibly old—owned by the Bishop family since 1760. The current owners (still part of the original family) grow several heritage varieties, with most going to a Halifax cider company called Tideview Cider. Tideview uses old apple varieties like Golden Russet, Baldwin, and Northern Spy to make its "distinctive" Nova Scotia-style cider with names such as Heritage, Heritage Dry, and Golden Russet.

I was able to connect with a heritage apple lover in Edmonton, Alberta. Gabor Botar is a horticultural technologist and amateur

OPPOSITE:
Gabor Bator's home orchard in Edmonton is set up to maintain germplasm and produce fruit for shows and home use, with no provision for the use of powered equipment. The trees, mostly apples and a few pears, are planted in a grid, only a few metres apart.
PHOTO: GABOR BOTAR

plant breeder who works at the University of Alberta. I found his name early in my research when I stumbled across an intriguingly titled article in the *Edmonton Journal*: "Edmonton a hotbed for heritage apples."[1] The story describes a mini-orchard located at the U of A's main research station, where Gabor is saving the germplasm of "a number of valuable clones so the plants can be reproduced." (Plant germplasm is the living tissue from which new plants can be grown. It contains the genetic information for the plant's hereditary makeup.) Gabor, who has lived in Edmonton most of his life, became interested in heritage apples with a summer job at the U of A's then Plant Science Department in 1977.

"Naturally, I tasted some of the fruit, particularly because we were using it as a teaching tool for a couple of courses, demonstrating storage qualities, et cetera. Anyway, once friends and neighbours had sampled some fruit, they were amazed to find such a range of flavours and textures, especially since most of them had assumed that one could grow only tough little crabapples here," he explained in an email. "I began observing unidentified fruit trees on boulevards and private gardens—I had a part-time pruning and grafting business for a couple of years after I graduated—and realized a lot of valuable and unique germplasm would be lost once people familiar with it moved on. As it is, one of my main concerns is trying to identify some old and not-so-old material correctly, a task that would best be tackled with DNA analysis."

His concern for trying to preserve germplasm took on greater urgency when the old U of A orchard was abandoned to government buildings in 1984, and then again in the early 1990s, when the much smaller reconstituted orchard at the station was terminated after a gradual decline.

"My wife and I moved here to the station in late 1990 and I immediately set about growing seedlings to use in grafting, with a view to saving what I could. Some of the cultivars in our garden here are from the remnants of the last orchard that I salvaged in 1994, by selecting scions from apparently disease-free branches. However, most of the original germplasm in the old orchard was lost."

Gabor also belongs to the Devonian Botanic Garden (DBG) Fruit Growers' Group, which organizes local scion exchanges and fruit shows. The fall fruit shows are particularly interesting, he said, as many people are unaware of the range of fruits that can be grown in Edmonton, and some return the following spring to take part in the scion exchange.

This brought me back to Edmonton's designation as a heritage apple "hotbed." In the *Journal* article, fruit tree expert Paul Olsen describes Edmonton as the "Apple Capital" of the prairie provinces, arguing the city has more apple trees and varieties of cultivars than any other. As well, it has the best apple-growing climate, good soil, and enough rainfall. Interestingly, he adds, "There are other factors, too, [such as] the eastern European influence—Ukrainians love apple trees, and I think they must have brought over some scionwood for grafting."

So what are Gabor Botar's favourite apples growing in Edmonton?

"It is gratifying that nurseries still carry some of the older and

A Miami apple, grown in Edmonton, is one of Gabor Botar's favourites. He says it is the single most-requested scionwood at the Devonian Botanic Garden Fruit Growers' Group scion exchanges.
PHOTO COURTESY GABOR BOTAR

more recent goodies like Goodland (1950s) [from Manitoba] and Norkent (1990)," he said. "I miss seeing more of Heyer 12 (1940), a standard for hardiness for many years and unusual for a real apple in that it bears most years and is good for both sauce and jelly. The best crabapple for jelly in my opinion is still the hardy and prolific Dolgo (1917); since it is one of the South Dakota originals, it may still be around in US nurseries as well is in Canada. A fair number of backyards in Edmonton still have this colourful-fruited cultivar. For sheer fresh flavour, I prefer the crabapple Rescue (1936) and the now rare but uniquely flavoured (Reinette-like) Harcourt (1955), a University of Alberta release. However, I am also partial to Miami (1959), at least the large-fruited strain I salvaged from the old U of A orchard. It is the single most-requested scionwood for me at the scion exchanges at DBG. Battleford (1934) and Luke (1961) are both large and good for sauce. For winter storage, my preference goes to Manitoba Spy (1931), which I think would substitute quite well for Granny Smith in some pastries. Prairie Spy (1940) is a similar good keeper."

There are other university-based orchards and germplasm repositories across the country, including the Heritage Apple Orchard at the University of Guelph, the successor to the Canadian Centennial Museum Orchard, established in 1967 to commemorate the one hundredth anniversary of Confederation in Canada. Today, Guelph's now-smaller heritage orchard features cultivars with commercial importance in Ontario's apple industry, past and present. Almost all of the cultivars formerly included in the centennial orchard are now maintained at the New York State Agricultural Experiment Station. This has been designated a national repository for apples by the US Department of Agriculture, and is the largest apple repository in the world.

In Canada, Agriculture and Agri-Food Canada preserves heritage fruit at its Greenhouse and Processing Crops Research Centre in Harrow, Ontario. The centre operates the largest greenhouse research facility in North America and is the home of the Canadian Clonal Genebank, which among other things collects, conserves, and distributes thirty-five hundred varieties of tree fruits and berries, including eight hundred types of apples.

I asked both Bob Weeden and Gabor Botar about the importance of individual orchardists saving heritage apple varieties via grafting and cultivation in light of the existence of germplasm repositories. Both stressed the significance of grassroots reproduction.

Bob said, "Yes, germplasm can be stored for a long time, then re-inserted into living plant (or animal) cell nuclei—pollen, ovary, sperm, or whatever—to produce another generation. Of course, there are some assumptions, such as that the germplasm is stored well and watched for thousands of years regardless of the political party in power, the cost of maintenance, et cetera. [Furthermore], plants and animals live in a habitat in an ever-changing world that they have to adapt to continually. Keeping a species going as a fixed form—let's say a McIntosh apple in a conservatory, grafted every century or so to replace old ones with young ones—is like unhooking a railroad car while the train moves on its journey. You might make the car and contents last a really long time, but they would become more and more irrelevant as the years passed."

Gabor stressed, "The only way to propagate fruit trees that produce a desired fruit with almost absolute certainty is via vegetative propagation," and that grafting is still the predominant method of propagating fruit trees while keeping the genetic lines stable.

The problem with repositories, he said, "is that the only practical way to store tree-fruit germplasm is in a living, grafted tree. A scion has a shelf life of only a few weeks under ideal refrigerated conditions. It is possible to store some germplasm long-term in a tissue-culture medium, but this tends to be expensive, and reconstituting the original plant often results in a number of off-types that must be carefully culled out. In short, yes, it is possible to maintain heritage-tree germplasm even once the original tree is gone, but this maintenance in most cases requires grafting and eventually more grafting; if this is not done conscientiously, the germplasm will be gone, forever."

The problem gets bigger on a political and financial level, Gabor said. Years ago, it was possible to order scions (first for free, then for a charge) from one of the government-run fruit germplasm repositories in Canada. Lately, though, this "cannot be done because the old orchards cost too much to maintain—and if they are not maintained, suckers, et cetera, make identification increasingly difficult and uncertain." Obsolete cultivars are removed, and only a few essential lines are kept, mainly for academic researchers.

"If a university or research station does breed new cultivars,

the latter are released to commercial orchards for propagation and eventual production. The commercial orchards have little interest in keeping old cultivars around when newer, more desirable ones (from a commercial viewpoint) can be grafted on top of their existing trees. Therefore, many of the old cultivars (which had flavours and textures modern commercial ones may lack) will literally disappear for good if members of groups, such as our DBG Fruit Growers, do not take it upon themselves to use their own time and resources to maintain desirable germplasm via scion exchanges—and of course grafting."

But he also sees problems with reliance on volunteers, who can't always be counted on in the long term because they die or move on. Also, there is the potential for mislabelling.

"I have seen as many as six different samples of fruit claiming to be the same cultivar—clearly, most or all have identity issues. Even commercial orchards mislabel some of their products: I have actually seen an apricot tree that turned out to be a plum, so it is not surprising that amateurs pass on mislabelled scionwood that they themselves originally propagated from mislabelled material."

To this end he has been proposing for years that a laboratory be established to undertake DNA analysis of heritage and modern fruits, and create a "realistic database of just what fruits we do have and which are endangered."

In the meantime, it will be up to the pockets of heritage apple-growing farmers and volunteer groups, from one end of Canada to the other, who will be saving rare varieties via community-based scionwood exchanges and grafting workshops—the Harry Burtons, Clay Whitneys, Bev Sidnicks, and Ron Schneiders, each quietly working away in their orchards, preserving small but essential pieces of living history.

Taste and appearance: Flat, round shape with a greenish-yellow skin and patches of brown russeting. Flesh is creamy-white with a sharp-sweet, unique, almost pear-like flavour.

Use: Eating fresh, juice, and hard cider.

History: Raised by Dr. Ashmead in Gloucester, England, around 1700.

Growing and harvesting: Picked in early October; the taste improves with age.

Other: Ashmead's Kernel is one of a very small number of English apple varieties that thrive in North America.

WEALTHY
MINNESOTA, 1860

Taste and appearance: Medium-sized, flat-round with pale yellow skin, splashed and striped with red. The flesh is creamy-white with red veins, crisp and juicy. It has a distinct, sprightly flavour.

Use: Eating fresh, cooking, and cider.

History: Raised in Minnesota in 1860 from an open-pollinated Cherry Crab seed obtained in Bangor, Maine.

Growing and harvesting: Picked in early September, it keeps until October, or November in cold storage.

Other: It remains a popular apple in Russia because it tolerates cold winters well.

PHOTO: DERRICK LUNDY

Taste and appearance: Medium-sized, conical with yellow and red skin. Deep, cream-coloured flesh with a fruity, lively taste. The flavour is said to improve with age.

Use: Eating fresh, cooking, and juice.

History: This highly respected American apple variety is named after the settlement of Esopus, Ulster County, New York, where it was found toward the end of the eighteenth century. It was rumoured to be Thomas Jefferson's favourite apple.

Growing and harvesting: Picked in mid-October, it keeps well in storage for up to five months. Once source says the tree is susceptible to scab; another says: "Susceptible to about all the common apple diseases."

Other: Jonathan is an offspring of this variety.

GOLDEN RUSSET
NEW YORK, 1800s

PHOTO: CLAY WHITNEY

Taste and appearance: Medium-sized, round, cylindrical in shape, with yellow skin and russeting. Flavour is sweet and crisp.

Use: Eating fresh, cooking, and cider.

History: The origins aren't clear but it arose in upstate New York in the nineteenth century, possibly derived from an English russet variety.

Growing and harvesting: Picked in October, it stores well until April.

Other: Considered to have the best flavour of American russet apples.

Apple-Spinach Tart

4 Tbsp olive oil

2 cups chopped leeks

12-oz bag fresh spinach, or 2 nice bunches,
    washed and stems removed

6 eggs

2 cups light cream or whole milk

1 tsp salt

1 tsp pepper

4 sheets phyllo pasty

2 cups blue cheese

6 apples (preferably Spartans and Kings)

2 egg whites, beaten

Sauté leeks in oil with a bit of salt until soft. Set aside. Slightly sauté spinach, get rid of extra water, and chop. Set aside. In another bowl, mix together eggs and cream with salt and pepper. To assemble the tart, line a 9 x 13-inch pan with phyllo sheets, oiling slightly between sheets. Crumble or distribute cheese over the pastry. Cover with leeks and spinach, and pour the egg mixture overtop. Quarter and core apples. Thinly slice each quarter into 8 or so pieces. Spread the apples decoratively across the top of the tart. Brush the top of the apples with egg whites. Cook at 350 degrees for 20 to 25 minutes, or until eggs are set. Don't burn the phyllo. Let the tart rest for 15 minutes before serving.

Recipe provided by Daphne Taylor, Salt Spring Island landscaper and caterer

Tarte Tatin

1 stick (½ cup) unsalted butter

1 egg, lightly beaten

2 Tbsp cold water

1 pinch salt

1⅔ cups all-purpose flour, sifted

Caramel Apple Filling:

½ stick (¼ cup) unsalted butter

¾ cup sugar

2 Tbsp water

10 apples

2 Tbsp sugar, for sprinkling

To make the pastry, let butter soften to room temperature and put it into a mixer (food processor) on low speed. Pulse for 2 seconds before adding egg followed by water. Mix for a few seconds and then add salt and flour, keeping 2 tablespoons aside to add later in case the dough is too sticky. Work fast because the gluten in the flour makes the dough go elastic. Stop the mixer before the dough turns into a ball. Flatten the pastry and shape into a circle about 6 inches wide. Place the pastry on a plate, wrap it in plastic wrap, and leave in the refrigerator for at least 1 hour or up to 24 hours. This lets the gluten relax and when you roll out the pastry it stays flat.

To make the caramel apple filling, cut butter into little bits and scatter over a 10-inch heatproof baking dish. Shake sugar over it and add 2 tablespoons of water to keep it from crystallizing. Caramelize the sugar by placing the dish on medium heat. Meanwhile, peel apples, cut in half, and remove the core.

Cut each apple into 4 big pieces. Once the butter and sugar have caramelized, take the pan off the heat. Place the pieces of apple vertically on top of the caramel in the baking tin, taking care to fill the gaps with more pieces so they stick together in a solid mass. Sprinkle sugar over the apples.

Put the dish back on medium heat for 15 minutes until the caramel starts bubbling up through the apples. Preheat the oven to 400 degrees. Roll the pastry out and place over the apples, folding it in at the edges. Make 3 or 4 holes with a knife and 1 in the middle to let steam out when baking. Bake for 20 minutes and then let it rest on the counter for 15 minutes. It is important that you do this, otherwise the apples will fall apart when you turn it over. Take a dinner plate and put it over the baking tin. Turn it over. Slowly remove the baking pan. Serve lukewarm with vanilla ice cream, heavy cream, or whipped cream.

Recipe provided by Steve Glavicich, chef/owner, Braizen Food Truck, Calgary

Conclusion

On Christmas morning 2011, Bruce led me to the back porch where he'd wrapped a cluster of three young Cox's Orange Pippin trees in a bright red ribbon. A few days later, clad in gumboots and rain gear, we tromped out back and amid the mild drizzle of a West Coast December, dug holes and planted the trees. It wasn't quite that simple, as every hole we shovelled immediately pooled with water and we spent most of the next hour digging trenches to drain the garden area. It felt like playing in the mud! Later, we collapsed inside, mud caked, refreshed from the tingle of wet air on our skin, and exuberant from "playing" in nature. A windstorm blew up in the night and first thing the next morning, we were up, outside, and inspecting the trees, regarded now with the nurturing reverence given to offspring.

I love the image of those fledgling trees in my backyard juxtaposed against the towering, grand old trees I've encountered over the last year. I get it. I understand the inspiration, the nostalgia, and the connection with our past and present to which these trees link us.

When I asked Jane Lighthall on Denman Island, "How and why did you become interested in heritage apples?" I was moved by the poetry of her response: "When [you] visit Denman Island you climb the hill from the ferry and take the main fork to the right where ancient apple orchards line the road. In winter, their big old bones stand regally and in season, the trees are often laden. The presence of these beauties certainly piqued my interest."

By growing heritage apple trees, Jane and her husband, Larry Lepore, are carrying on a family tradition—Jane's grandparents were pioneers in Oliver, BC, in the late 1920s, and began an orcharding tradition that carried on into the next generation.

"Unbeknownst to us, the land that we bought bordered on what was once an orchard of one thousand trees planted at the turn of the last century by a pioneering family, and many of these trees are still producing. Denman has several such original orchards with trees over a hundred years old that have truly become a legacy and an inspiration."

Their customers come from "all walks of life"—apple enthusiasts, some "just discovering the old varieties, [others who] have long been growing apples and just need a few more trees to complete their orchards." Some want a tree or two for their backyard, "but more and more people want to grow their own food."

And this is how the apple trees of our past will feature in our future. This is why heritage apples will slowly, steadily re-emerge.

The author, Susan Lundy, with one of her recently planted Cox's Orange Pippin trees in the spring of 2012.
PHOTO: DANICA LUNDY

We will rediscover heritage apples at festivals, farm stands, markets, and small, real-food grocery stores. Through the efforts of people like Clay Whitney, Gabor Botar, and the Brae Island Park Board we will recognize the living links—the very deep roots—to our past that tower majestically above us, sometimes lost amid the rampant chaos of overgrown orchards, other times threatened by urban development. Slowly, we will learn that the beauty of an apple is more than skin deep. We will bite into oddly shaped orbs, maybe with brown russeting on their shoulders, maybe small, green, and twisted—not big, red, and shiny—and discover that taste is actually much more than an undressing of our eyes. Red-fleshed apples will move beyond a chef-driven novelty and into our homes as we revel in their flavour and health benefits. And as we rediscover cider, a beverage that has much more structure and excitement than the common drink found cooling in cans at the liquor store, the trees of our past will grow in greater significance.

But ultimately, heritage apples will morph into "today's apples" because we care more and more about where our food comes from, who is growing it, and how far it has travelled to get to our plates. The apples of our past will become the apples of our eyes.

Back at home, spring approached and we eagerly awaited the first bud that said our trees were thriving. Someday, I will pluck my first homegrown Cox's Orange Pippin. Maybe, just maybe, I will use them to bake my first apple pie. But for sure, I will join the ranks of apple people hunting down new flavours from old varieties. I have that gleam in my eye. I am an apple person.

Endnotes

Chapter One: The Apples of My Eye

[1] Shane Peacock, *Mr. McIntosh's Wonderful Apple*, http://sgtmajmac.tripod.com/
macapple.html.

[2] "Food Movement, Rising," *The New York Times Review of Books*, May 20, 2010.

[3] *Ibid.*

[4] The dates of apple origins sometimes vary by a few years according to the source.
In most cases, I have used dates supplied in the apple variety index of the website Orange Pippin, http://www.orangepippin.com.

Chapter Two: "A" is for Apple

[1] *Wikipedia*, http://en.wikipedia.org/wiki/Apple.

[2] The University of Illinois website, http://urbanext.illinois.edu/apples/facts.cfm.

[3] *Wikipedia*, http://en.wikipedia.org/wiki/Apple.

[4] Jim Rahe, "Singing the Praises of Gravenstein," *The Cider Press,* BC Fruit Testers
Association archives.

[5] Michael Pollan, *The Botany of Desire: A Plant's-Eye View of the World* (New
York, Random House, 2001), 11.

[6] LifeCycles Project Society, http://lifecyclesproject.ca.

Chapter Three: From the Garden of Eden

[1] Midwest Apple Improvement Association, http://www.hort.purdue.edu/newcrop/
maia/history.html.

[2] *The Precious Book of Enrichment* is an ancient Chinese text, apparently written
around 5000 BC.

[3] Henry David Thoreau, "Wild Apples," *The Atlantic*, 1862.

4   Dr. Barrie Juniper and David J. Mabberley, *The Story of the Apple* (Portland, Oregon, Timber Press, 2006).

5   Dr. Barrie Juniper, "The Mysterious Origin of the Sweet Apple," *American Scientist*, January–February, 2007.

6   Pollan, *The Botany of Desire*, 7.

7   Pollan, *op cit*, 9.

8   Pollan, *op cit*, 23.

9   Carol Martin, *The Apple: A History of Canada's Perfect Fruit* (Toronto, Ontario, McArthur & Company, 2006).

10 *Ibid.*

11 Apple Luscious Organic Orchard, http://www.appleluscious.com.

12 "The Islands—A Well Known Fruit District," Victoria *Daily Colonist*, December 13, 1908.

13 *The Garden of Eden: The History of Apple Orchards in the Okanagan Valley*, http://www.nald.ca/library/learning/okanagan/history/6garden.pdf.

Chapter Four: How Do You Like Them Apples?

1   The University of Illinois, http://urbanext.illinois.edu/apples/facts.cfm.

2   *The Apples of the Apple Festival*, pamphlet written and produced by The Friends of the UBC Botanical Garden.

Chapter Five: Biting Into a Surprise

1   Orange Pippin, http://www.orangepippin.com.

2   *The Fruit Gardener*, published by the California Rare Fruit Growers, Volume 27, Number 3, May/June 1995.

3   Global Trees Campaign, http://www.globaltrees.org/kyrgyzstan_apple.html.

4   Richard Espley, "Under the Skin of the Red-Fleshed Apple," http://www.youtube.com/watch?v=w4uGw48OYDI.

Chapter Six: "Surely the Noblest of all Fruit"

1   The University of Illinois, http://urbanext.illinois.edu/apples/facts.cfm.

2   "Communities In Bloom," British Columbia, Evaluation Form, 2011, http://www.sooke.ca/assets/What/Sooke2011Report.pdf.

3   "Who We Are," BC Fruit Testers Association, http://bcfta.ca.

4   Seeds of Diversity, http://www.seeds.ca/en.php.

5   Creemore Heritage Apple Society, http://creemoreheritageapples.ca.

6   *Ibid.*

7   Orange Pippin, http://www.orangepippin.com.

8   *An Apple a Day*, Mary Mollet, self-published, 1995.

Chapter Seven: "Comfort Me with Apples"

1   *The Acorn*, Salt Spring Conservancy, Number 36, Fall 2007, http://www.saltspringconservancy.ca/newsletter.html.

Chapter Eight: Deep Roots

1   "In The Classroom Facts and Folklore," BC Agriculture, http://www.sfvnp.ca/sg_userfiles/APPLES_2011.pdf.

2   Jessie Ann Smith, as told to J. Meryle Campbell and Audrey Wild, *Widow Smith of Spences Bridge* (Merritt, British Columbia, Sonotek Publishing, 1989), 9.

3   Hoodoo Ranch, http://www.hoodooranch.ca.

Chapter Nine: The Cider House Rules

1   History of Cider Making, http://www.chm.bris.ac.uk/webprojects2003/lim/Appleweb2003/history.html.

2   Definition according to the *Merriam-Webster Dictionary.*

3   Merridale Estate Cidery, http://www.merridalecider.com.

Chapter Ten: The Apple Doesn't Fall Far from the Tree

1   M. Maclean, "Edmonton a hotbed for heritage apples," *Edmonton Journal*, September 7, 2007.

Index

Acknowledgments

My greatest thanks to the many people who shared their time, enthusiasm, extensive knowledge—and their apples—as I researched this book. Dr. Bob Weeden answered many questions, responded to numerous emails, and ultimately read the entire manuscript for me. It was great to have his careful eye peruse these chapters. Thanks also to Harry Burton and Clay Whitney, who, several times, let me follow them around asking question, and read and responded to chapters in the book; and to Bev Sidnick, Ron Schneider, Kristen Jordan, Gabor Botar, Bill Wilde, and Janaki Larsen, who also gave me their time. Thanks to Steve Glavicich, Daphne Taylor, Bev Sidnick, Marjorie Lane, and Mary Mollet for the recipes and Judy Weeden for the cooking tips. Thanks also to Derrick Lundy, Sierra Lundy, Danica Lundy, Clay Whitney, and Harry Burton for the photos. A massive thanks to Richard Borrie at Orange Pippin Ltd. and Hamid Habibi at Keepers Nursery, who came through at the last minute with some beautiful photographs of apples I simply could not find anywhere else. And, of course, thanks to Bruce Cameron for the endless enthusiasm, the apple drinks, treats, and products, and ultimately, all the support.

SUSAN LUNDY is an award-winning journalist, editor, and freelance writer. She is a two-time recipient of the prestigious Jack Webster Award of Distinction and has won more than 25 writing, design, and project awards. After working for many years at the *Gulf Islands Driftwood*, she now focuses on freelance writing, specializing in travel and personal narrative. She has a degree in creative writing and journalism from the University of Victoria, is the editor of *Soar* and *Tweed* magazines, and writes a column on family life for several newspapers. Susan divides her time between Victoria and Salt Spring Island, British Columbia, and Calgary, Alberta.